Handbook of Petroleum Analysis

Handbook of Petroleum Analysis

Editor

Jyoti Sharma

Handbook of Petroleum Analysis
Edited by **Jyoti Sharma**

Printed in 2017

ISBN: 978-1-68117-395-5

Library of Congress Control Number: 2015941585

© 2016 by
SCITUS Academics LLC,
616, Corporate Way, Suite 2, 4766,
Valley Cottage, NY 10989

www.scitusacademics.com

This book contains information obtained from highly regarded resources. Copyright for individual articles remains with the authors as indicated. All chapters are distributed under the terms of the Creative Commons Attribution License, which permits unrestricted use, distribution, and reproduction in any medium, provided the original author and source are credited.

Notice

Reasonable efforts have been made to publish reliable data and views articulated in the chapters are those of the individual contributors, and not necessarily those of the editors or publishers. Editors or publishers are not responsible for the accuracy of the information in the published chapters or consequences of their use. The publisher believes no responsibility for any damage or grievance to the persons or property arising out of the use of any materials, instructions, methods or thoughts in the book. The editors and the publisher have attempted to trace the copyright holders of all material reproduced in this publication and apologize to copyright holders if permission has not been obtained. If any copyright holder has not been acknowledged, please write to us so we may rectify.

Contents

Preface vii

Chapter 1 Synthesis of Petroleum-Based Fuel from Waste Plastics and
Performance Analysis in a CI Engine 1
Christine Cleetus, Shijo Thomas, and
Soney Varghese

Chapter 2 Molecular Components-based
Representation of Petroleum Fractions 27
Muhammad Imran Ahmad, Nan Zhang, and
Megan Jobson

Chapter 3 Molecular Reconstruction of LCO Gasoils from
Overall Petroleum Analyses 61
Damien Hudebine and Jan J. Verstraete

Chapter 4 Experimental Campaign, Modeling, and Sensitivity
Analysis for the Molecular Distillation of Petroleum
Residues 673.15K+ 83
L. Zuñiga Liñan, N.M. Nascimento Lima, F. Manenti,
M.R. Wolf Maciel, R. Maciel Filho, and L.C. Medina

Chapter 5 Removal of Petroleum Sulfonate from Aqueous Solution
by Hydroxide Precipitates Generated from Leaching
Solution of White Mud 133
Lu Cheng, Lanlan Ye, Dejun Sun, Tao Wu, and Yujiang Li

Chapter 6 Electrochemical Treatment of Effluents from
Petroleum Industry Using a Ti/RuO2 Anode 167
Iranildes D. Santos, Márcia Dezotti, and Achilles J.B. Dutra

Chapter 7	Analysis of Petroleum-contaminated Soils by Diffuse Reflectance Spectroscopy and Sequential Ultrasonic Solvent Extraction–gas Chromatography 191
	Reuben N. Okparanma, Frederic Coulon, and Abdul M. Mouazen
Chapter 8	Molecular Reconstruction of Heavy Petroleum Residue Fractions 221
	J.J. Verstraete, Ph. Schnongs, H. Dulot, and D. Hudebine

Citations 249

Index 253

Preface

Petroleum exhibits a wide range of physical properties. Numerous tests have been and continue to be developed to provide an indication of the means by which a particular feedstock should be processed. An initial inspection of the nature of petroleum provides deductions about the most logical means of refining and classifying. Handbook of Petroleum Analysis is a single, comprehensive source that describes the application and interpretation of data resulting from various test methods for petroleum feed stocks and products. Thus this book deals with the various aspects of petroleum analysis and provides a detailed explanation of the necessary standard tests and procedures that are applicable to feed stocks to help define predictability of behavior. In addition, the application of new methods for determining instability and incompatibility as well as analytical methods related to environmental regulations are described.

Editor

Chapter 1

Synthesis of Petroleum-Based Fuel from Waste Plastics and Performance Analysis in a CI Engine

Christine Cleetus[1], Shijo Thomas[2], and Soney Varghese[2]

[1]Department of Mechanical Engineering, NIT Calicut, Kerala 673601, India
[2]School of Nano Science and Technology, NIT Calicut, Kerala 673601, India

ABSTRACT

The present work involves the synthesis of a petroleum-based fuel by the catalytic pyrolysis of waste plastics. Catalytic pyrolysis involves the degradation of the polymeric materials by heating them in the absence of oxygen and in the presence of a catalyst. In the present study different oil samples are produced using different catalysts under different reaction conditions from waste plastics.

The synthesized oil samples are subjected to a parametric study based on the oil yield, selectivity of the oil, fuel properties, and reaction temperature. Depending on the results from the above study, an optimization of the catalyst and reaction conditions was done. Gas chromatography-mass spectrometry of the selected optimized sample was done to find out its chemical composition. Finally, performance analysis of the selected oil sample was carried out on a compression ignition (CI) engine. Polythene bags are selected as the source of waste plastics. The catalysts used for the study include silica, alumina, Y zeolite, barium carbonate, zeolite, and their combinations. The pyrolysis reaction was carried at polymer to catalyst ratio of 10 : 1. The reaction temperature ranges between 400°C and 550°C. The inert atmosphere for the pyrolysis was provided by using nitrogen as a carrier gas.

INTRODUCTION

In the recent years it is quite common to find in newspapers and publications that plastics are turning out to be a menace. Days are not so far when earth will be completely covered with plastics and humans will be living over it. All the reasoning and arguments for and against plastics finally land up on the fact that plastics are nonbiodegradable in nature. The disposal and decomposition of plastics has been an issue which has caused a number of research works to be carried out in this regard. Currently the disposal methods employed are land filling, mechanical recycling, biological recycling, thermal recycling, and chemical recycling. Of these methods, chemical recycling is a research field which is gaining much interest recently, as it turns out to be that the products formed in this method are highly advantageous.

Plastic is one such commodity that has been so extensively used and is sometimes referred to as one of the greatest innovations of the millennium. There are a numerous ways in which plastic is and will continue to be used. The plastic has achieved such an extensive market due to fact that it is lightweight, cheap, flexible, reusable, do not rust or rot, and so forth. Because of this, plastics production has gone up by almost 10% every year on a global basis since 1950 [1]. Asia accounts for 36.5% of the global consumption and has been world's largest plastics consumer for several years. The major segment continues to be the packaging, which has accounted for over 35% of the global demand [2]. The global per capita consumption of plastics is shown in Figure 1.

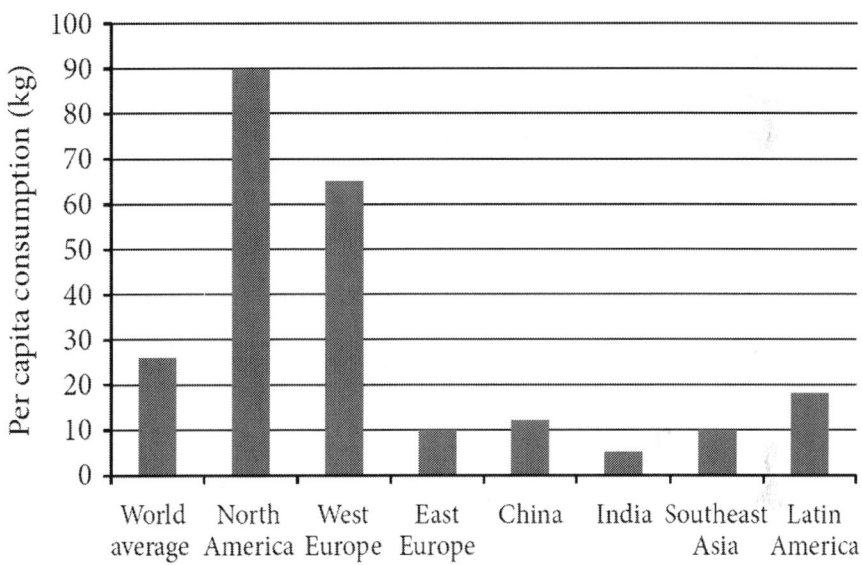

Figure 1: Global per capita consumption of plastics [2].

The global production of plastics has seen an increase from around 1.3 million tonnes in 1950 to 245 MT in 2006 [1]. In recent years, significant growth in the consumption of plastic globally has been due to the introduction of plastics into newer application areas such as in automotive field, rail, transport, aerospace, medical and healthcare, electrical and electronics, telecommunication, building and infrastructure, and furniture. This significant growth in the demand for plastic and its forecast for future have certainly proved that there has been a quiet plastic revolution taking place in every sector.

As far as the individual plastics materials are concerned, polyolefins account for 53% of the total consumption. The consumption of the individual plastic materials is shown in the Table 1. It can be seen that one-third of the global consumption of plastic is polythene. The growth in the global polythene demand is estimated to be around 4.4% annually up to 2020 [2]. This is the reason behind the selection of polythene as the source of waste plastic in this study.

Table 1: Global consumption of individual plastics [2]

Type of plastic	Consumption %
Polythene (PE)	33.5
Polypropylene (PP)	19.5
Polyvinylchloride (PVC)	16.5
Polystyrene (PS)	8.5
Polyethylene terephthalate (PET) and polyurethane (PU)	5.5
Styrene copolymers (ABS, SAN, etc.)	3.5
Blends, alloys, high performance and specialty plastics, thermosetting plastics, and so forth	13

The increase in the rate of plastic consumption throughout the world has led to the creation of more and more amounts of waste, and this in turn poses greater difficulties for disposal. This is because the life duration of plastic (the time period for which the plastic remains in use) is very small. About 40% of plastics consumed have duration of life smaller than one month. The service life of plastic products ranges from 1 to 35 years depending on the area of application. In India, the weighted average service life of all plastics products comes to about 8 years. This may vary among countries depending on the type of consumption. This short service life in India reflects that a major share of the plastic consumed here is short-life products. This can be accounted for as the share of plastics used in packaging which is 42% [1].

DIFFERENT METHODS OF PLASTIC WASTE MANAGEMENT

The suitable treatment of plastic is the most important factor in waste plastic management. This is quite important from the energetic, environmental, and economic point of view. Even though the recycling rate for postconsumer plastics has increased in the recent years, this increase has been only meager coming to only around 1.5%. This increase in the recycling is due to the strict legal regulations and growing awareness. Different techniques for the waste plastic management are being followed today [1].

The major portion of the waste plastics has been subjected to landfill. Such a disposal of the waste to landfill is now strictly regulated legally. The regulations are expected to achieve a reduction of 35% in land filling over the period from 1995 to 2020. Also the rising cost and scarcity of land, the generation of explosive greenhouse gases (such as methane), a high volume to weight ratio of plastics, and the poor biodegradability of commonly used packaging polymers also make it an unattractive option.

Reprocessing of the used plastics to form new similar products is termed as mechanical recycling. In this method, the products obtained are with almost same or less performance level than the original product. Even if the technique seems to be "a green operation," it is not cost effective as it requires high energy for cleaning, sorting, transportation, and processing to make a serviceable product. Practically it is seen that reprocessing of mixed contaminated plastics yields mechanically inferior products lacking in durability compared with the original polymers.

Biodegradable polymers are those which can be converted back to the biomass in a realistic time period. However, there are a number of difficulties over the use of degradable plastics. First, appropriate conditions are necessary for the degradation of such plastics, such as presence of light for the photodegradable plastics. Second, greenhouse gases such as methane is released when plastics degrade anaerobically. This is by enabling microorganisms in the environment to metabolize the molecular structure of plastic films to produce an inert humus-like material that is less harmful to the environment.

Incineration of plastics waste is an alternate method in which energy is recovered from waste polymers. These hydrocarbon polymers can replace fossil fuels and thereby reduce the CO_2 burden on the environment. Polyethylene is having calorific value similar to that of the fuel oil, and the energy produced by incineration of polyethylene is of the same order as that used in its manufacture making it an attractive option. However, this method produces greenhouse gases and toxic pollutants giving it a big disadvantage.

Cracking process breaks down the long polymeric chains into useful smaller molecular weight compounds. The products of this process are highly useful and can be utilized as fuels or chemicals in various applications. The pyrolysis reaction can be carried out without or in the presence of a catalyst. If without catalyst, it is thermal cracking or thermolysis, and if in the presence of catalyst, it is catalytic pyrolysis.

Thermal cracking or pyrolysis involves the degradation or cracking of the polymeric materials by heating them to a very high temperature. The heating should be carried out in the absence of oxygen to make sure that no oxidation of the polymer takes place. The temperature ranges between 350 and 900°C. The products formed include a carbonized char (solid residues) and a volatile fraction. A portion of the volatile fraction can be condensed to give paraffins, isoparaffins, olefins, naphthenes, and aromatics, while the remaining is a noncondensable high calorific value gas. The products formed and their precise composition depend on the type of the plastic waste and the process conditions. In catalytic cracking, the same process is carried out in the presence of a catalyst. The prominent advantage of this method is that the presence of catalyst lowers the reaction temperature and time. Another added advantage is that in thermal cracking a broad variety of products are formed by the braking of the polymeric chain, while in catalytic degradation the product distribution will be a much narrower, with a peak at lighter hydrocarbons. From the economic point of view also, reducing the cost even further will make this process more attractive. Due to these reasons in the present work, this method is adopted for the synthesis of petroleum-based oil. The importance of this work lies in comparing the performance of different catalysts like barium carbonate, zeolite 1 (pore size ~4 Å), silica alumina 1 (SA1) (silica (~30 nm) 83.3%, alumina (~30 nm) 16.7%), silica alumina 2 (SA2) (silica 21.1%, alumina 78.9%), SA1 + Z1 (70% SA1, 30% Z1), and zeolite 2 (sodium Y zeolite) for the thermal cracking of waste polythene and selecting the most suitable catalyst based on the yield and thermophysical properties of the hydrocarbon oil obtained.

LITERATURE REVIEW

A lot of research has been done in noncatalytic and catalytic pyrolyses of plastics which proves that plastics waste can indeed be converted to useful chemical feedstock. The works reveal that the product distribution can be affected by a number of parameters which include the polymer source (plastic type), catalysts used, size of the catalyst, catalyst to polymer ratio, reaction temperature, reaction time, and reactor type. The effect of various process variables is described below.

Effect of Catalyst

The catalyst used in the pyrolysis of plastics definitely influences the product. The most commonly used catalysts in the literature for plastic waste pyrolysis includes silica alumina, zeolites (beta, USY, ZSM-5, REY, clinoptilolite, etc.), and MCM-41. With increasing number of acid sites, the level of the catalyst activity in polyolefin pyrolysis also increases. Thus, zeolite-based catalysts due to their high acid strength achieve higher conversion than nonzeolitic catalysts. Songip et al. [3] studied the conversion of polyethylene to transportation fuel using HY, rare earth metal-exchanged Y-type (REY), and HZSM-5 zeolites and silica-alumina (SA). It was found that REY zeolite was the most suitable catalyst producing plastic oil with the highest octane number and gasoline yield. REY had large pores and had proper acidic strength which made it the most suitable one. Y zeolite and ZSM-5 zeolite produced oils having a high research octane number comparable to that of the oil by REY, but the gasoline yield by the formers was significantly low as compared to REY. The catalytic degradation of polyethylene by ultrastable-Y zeolite was studied by Manos et al. [4]. Low pyrolysis temperature does not cause the polymer to fully degrade, and a solid residue is produced in the reaction bed. It showed that the catalyst has significantly reduced the degradation temperature as compared with pure thermal degradation in the absence of a catalyst. The products of the catalytic degradation were hydrocarbons in the C_3–C_{15} range. The catalyst was highly acidic, producing oil with high octane number.

A number of research works have been done to find out the effect of silica alumina as catalyst for the pyrolysis of plastics. It can be seen that with silica alumina, high liquid yield can be obtained. The effects of silica alumina with two different SiO_2/Al_2O_3 proportions; that is, SA-1 (SiO_2/Al_2O_3 ratio of 83.3/16.7) and silica alumina SA-2 (SiO_2/Al_2O_3 = 21.1/78.9) were studied by Uddin et al. [5]. The liquid yield was found to be 68 wt% for SA-1 as compared to 77 wt% for SA-2. Therefore, the SA-1 catalyst degraded the polyethylene sample into much lighter hydrocarbon fuel oil than the SA-2 catalyst. Thus it can be concluded that the yield and composition of the liquid products can be controlled by altering the SiO_2/Al_2O_3 ratio. The liquid products are distributed in C_5–C_{20} range, that is, basically in the gasoline and diesel ranges. The effect of nonacidic catalysts for the pyrolysis of plastics was studied by Jan et al. [6]. On comparison with $MgCO_3$ when used as a catalyst under 450°C, it could be observed that the % oil yield (33.60%) is higher with $MgCO_3$ as compared to the % oil yield (29.60%) obtained with $BaCO_3$ catalyst. Similarly when $CaCO_3$ was used as a catalyst under the same reaction conditions, the obtained % oil yield was 32.20%.

Effect of Catalyst Contact Mode

There exist two methods by which catalyst can be added to the pyrolysis reactor: liquid phase contact and vapor phase contact. In the liquid phase contact, the catalyst and polymer are mixed together, and then they are placed in the reactor and heated to the reaction temperature. However, in the vapor phase contact, the polymer is first subjected to thermolysis to produce the volatile fraction. The catalyst is inserted in the path of the moving vapour, and as the vapour moves through the catalyst, the hydrocarbon vapour is degraded to get the required product distribution. However, the product yield is reported not to differ significantly with the two modes [1].

Effect of Polymer to Catalyst Ratio

Effect of polymer to catalyst ratio has been studied by Akpanudoh et al. [7]. It has been concluded that with the increase in the amount of catalyst, a direct proportionality in terms of the effectiveness is not obtained. The increase in catalyst amount increases the conversion up to particular limit, but a further increase in the catalyst percentage does not give any appreciable increase in the conversion rate. The optimum polymer to catalyst ratio as obtained from studies is 4:1. However, it is also found in the literature that a lesser catalyst ratio will also provide similar degradation, but only at higher reaction temperatures [8]. Some kind of optimisation has to be done with the catalyst ratio and temperature, so that the operation remains economical too.

Effect of Temperature

If the catalytic pyrolysis is taking place at higher operating temperature or at high heating rates, it causes the enhancement of bond breaking and thereby favouring the production of smaller molecules. The extend of conversion increases with increase of temperature, and it can be seen that with higher conversion the major products formed will be the gaseous products and the liquid yield being minimum or nil. The effect of different catalysts on the liquid yield and the product distribution becomes less significant with increasing temperature. The reaction taking place will be similar to thermal degradation [8].

Effect of Flow Rate of Nitrogen Gas

The inert gas flowing through the reaction does not affect the reaction directly, but it can produce a slight change in the liquid yield. Usually the nitrogen flow rate was chosen to be relatively high, in order to move the volatile primary products from the reactor and keep secondary reactions at a minimum. This actually favours the liquid yield. But studies of Gulab et al. [8] indicate that high carrier gas flow rate can enhance the evaporation of liquid products which are collected in the condenser. This falsifies the results of liquid yield. By course of experiments, it has been found that the optimum flow rate is 10 mL/min.

The objective of the present work is to synthesize petroleum based fuel by the catalytic pyrolysis of waste plastics using different catalysts, optimization of the yield based on catalyst and reaction conditions, and its performance analysis in an IC engine.

WASTE PLASTIC OIL PRODUCTION

Polythene is selected as the source of waste plastics since it comprises a prominent percentage of the waste plastic produced. The catalysts identified for the study include silica alumina, zeolites, barium carbonate, titanium chloride, and their combinations. The pyrolysis reaction is carried out in the polymer to catalyst ratio of 4 : 1. The reaction temperature ranges between 350 and 450°C. The inert atmosphere for the pyrolysis is provided by using nitrogen as a carrier gas, and the flow rate is fixed to be 10 mL/min. These selections have been made on the basis of the literature survey that has been carried out. A schematic sketch of the experimental setup is as shown in Figure 2. The setup consists of a ceramic electric heater (reactor), a steel container, two condensers, two oil collectors, and a nitrogen source (inert gas). The maximum loading capacity of the reactor was 1.5 kg of waste plastics. The reactor consists of three ceramic heaters, each having power of 2000 W, arranged in series over a cast iron pipe of 17 cm diameter and 60 cm length. The double pipe counter flow heat exchanger of length 90 cm functions as the condenser. Water at around 28°C (room temperature) is used as the coolant in the first condenser, and the temperature of water being supplied to the second heat exchanger is 10°C. The waste plastic is placed inside a steel container of 15 cm diameter and 20 cm height at a packing density of 424 kg/m³. This container is finally placed inside the heater. The purpose of this container is to avoid the flow of melted waste plastic downwards under gravity as such a flow can block the passage of the nitrogen gas.

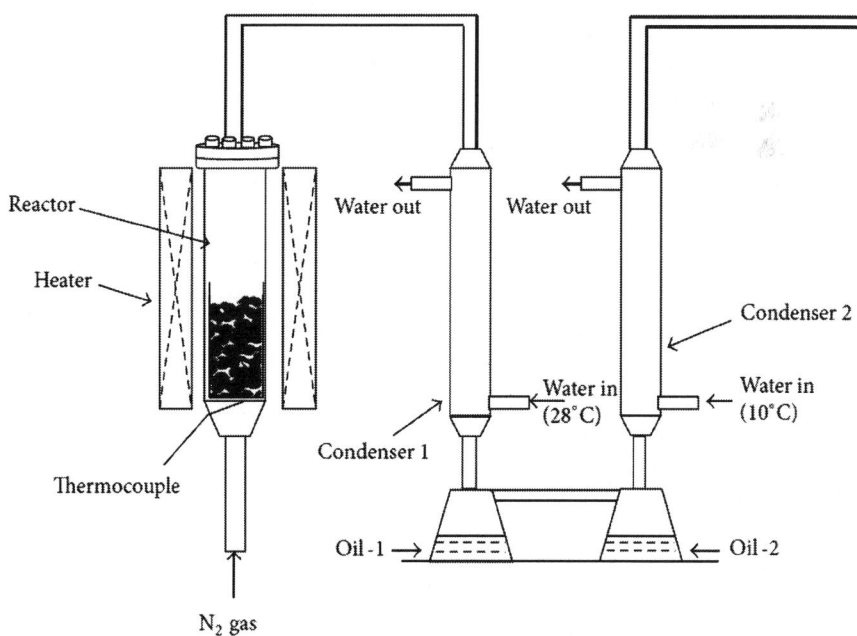

Figure 2: Schematic model of the experimental setup.

The waste plastic was mixed with the catalyst in a twin roll mill, before it is being supplied to the heater. During the mixing process the plastic gets heated enough to get devoid of any moisture content. The reactor was fixed vertically, and nitrogen gas was introduced into the reactor from the bottom. The flow of nitrogen replaces the air from the reactor and permits the pyrolysis reaction under anaerobic condition. Before starting the heating, nitrogen gas is allowed to flow through the heater unit to remove the oxygen that is present initially. Then heater is switched on, and the temperature controller is set to the required operational temperature. The vapor fraction formed during the pyrolysis of the plastic inside the reactor flows out along with N_2 out of the reactor. The gas mixture is first cooled in the condenser 1. Ordinary room temperature water is supplied to the condenser for the cooling action. The low boiling fractions of the vapour fraction will be condensed and collected in a collector fitted to the condenser 1. The remaining uncondensed fraction moves to the condenser 2. The cooling water for condenser is at a temperature of 10°C. This low temperature for water is provided by an external refrigeration setup. The low boiling components moving through condenser 2 will be condensed and collected in a collector. The remaining uncondensed part escapes to the atmosphere.

The oil samples are produced without using catalysts and also in the presence of catalysts such as barium carbonate, zeolite 1 (pore size ~4 Å), silica alumina 1 (SA1) (silica (~30 nm) 83.3%, alumina (~30 nm) 16.7%), silica alumina 2 (SA2) (silica 21.1%, alumina 78.9%), SA1 + Z1 (70% SA1, 30% Z1), and zeolite 2 (sodium Y zeolite), at reaction temperatures ranging from 400°C to 550°C.

The polythene waste is first shredded to sizes of 1-2 cm and then mixed with the test catalyst (polymer to catalyst ratio of 10:1) uniformly in the two roll mill with friction ratio of 1:1.4. The mixed plastic and catalyst are then inserted into the heater assembly and are heated to the required temperature of approximately 400°C. An inert gas (nitrogen) is passed through the heater assembly so as to prevent any kind of oxidation reaction that may take place. Once the temperature is attained, it is maintained for a preset reaction time (say 3 hours). After the reaction time is over, the heater is switched off, but the reaction is allowed to take place for another one more hour, so that maximum volatile fraction formed will pass through the two heat exchangers. The condensed volatile fraction is finally collected from the collecting tanks and is filtered. Now the process is repeated for different temperatures for various catalysts.

YIELD OF BIO-OIL AND OIL PROPERTIES

Out of the oil samples produced by different catalysts at different temperatures, the samples which showed optimum liquid yield for a particular catalyst are shown along with their properties in Table 2. It was noted that the liquid yield was available only at temperatures above 350°C for all catalysts. With the increase in temperature, the liquid yield decreased after a particular temperature (which was different for different catalyst). Finally, above 600°C no liquid yield was obtained for any catalyst thereby making the liquid yield gate between 400 and 550°C. Along with the oil obtained, a gel portion was also present along with the impurities which was filtered and removed. From the obtained results of liquid yield and its properties, the catalyst combination of silica alumina 1 and zeolite 1 (70% SA1, 30% Z1) is selected as the best catalyst with the optimum reaction temperature as 425°C. The oil produced using this optimum catalyst was used for GC-MS and was produced in adequate amount for the preparation of blends for testing in the engine.

Table 2: Optimum properties of plastic oil with different catalysts

Catalyst	None	$BaCO_3$	Z1	SA1	SA2	SA1 + Z1	Z2
Temp. (°C)	450	450	450	425	450	425	450
Liquid collected (mL)	80	50	110	145	130	130	100
After filtration (mL)	60	40	100	130	120	120	90

Calorific value (MJ/kg)	41.35	36.61	45.15	41.36	36.83	44.57	42.24
Viscosity (10^{-3} Ns/m²)	1.2699	1.6534	1.1891	1.2865	1.1995	1.2066	1.2245
Density (g/cc)	0.8581	0.9745	0.8635	0.9471	0.9106	0.9265	0.8785
Flash point (°C)	<32	—	<32	<32	<32	<32	<32
Fire point (°C)	≤32	—	36	34	36	35	34
Cloud point (°C)	−3	—	−2	−3	−3	−2	−2
Pour point (°C)	−13	—	−12	−12	−13	−13	−12

CHEMICAL COMPOSITION OF THE OIL SAMPLE

Gas chromatography-mass spectrometry (GC-MS) of the selected oil sample was done to examine its chemical composition. Chloroform was used as the solvent for the oil sample. The mass spectrum shows a number of peaks showing that the oil is a mixture of a number of compounds. The compounds vary from C_9 to C_{19}, showing clear similarities to diesel in addition to the oil properties showing closeness to that of diesel. The mass spectrum is shown in Figure 3.

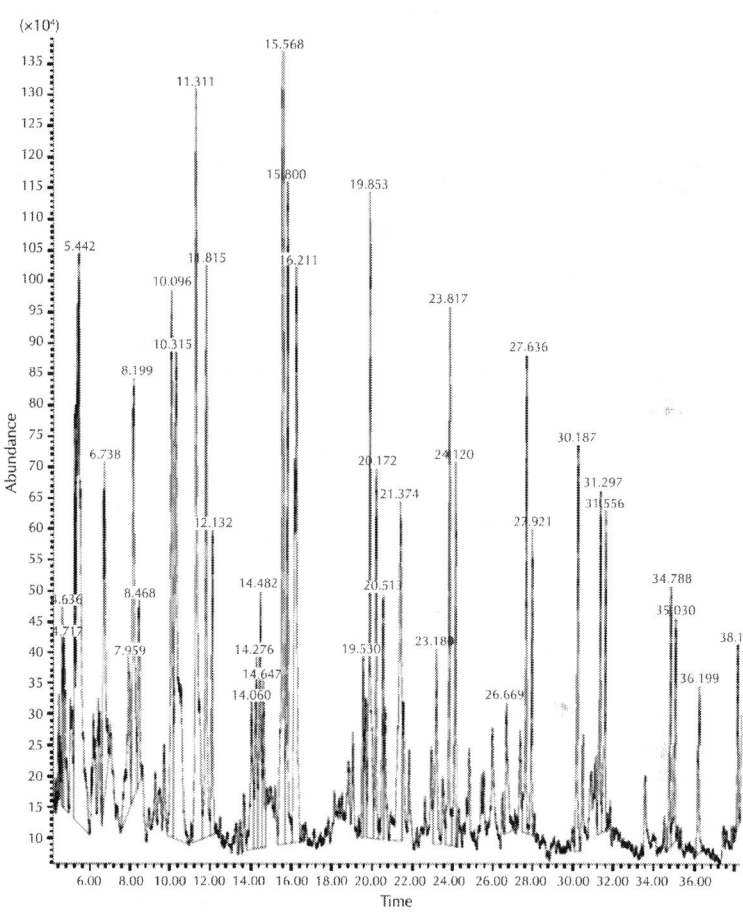

Figure 3: Mass spectrum of plastic oil sample (GC-MS).

The list of compounds shown in the mass spectrum, along with the area % is listed in the Table 3. From the area %, the average chemical formula of the sample oil can be calculated and is obtained as $C_{13.18}H_{23.56}$ (the average chemical formula of diesel is $C_{12}H_{23}$). The equation for the complete combustion of the plastic oil is as follows:

$$C_{13.8}H_{23.56} + 19.07 O_2 + 19.07\,(3.76 N_2)$$
$$\longrightarrow 13.18 CO_2 + 11.78 H_2O + 71.7 N_2 \qquad (1)$$

The theoretical air fuel ratio of the plastic oil is found to be 14.47. Due to the close similarities of the properties of the oil with diesel, the performance test of the oil is done in a CI engine.

Table 3: Chemical composition of plastic oil sample

Peak	Retention time (min.)	Composition	Chemical formula	Area %
4	6.738	Indene	C_9H_8	6.28
5	7.96	2-Butenyl-benzene	$C_{10}H_{12}$	3.94
6	8.201	1-Undecene	$C_{11}H_{22}$	7.39
11	11.817	1-Dodecene	$C_{12}H_{24}$	7.88
12	12.314	Dodecane	$C_{12}H_{26}$	3.39
13	14.062	1,3-Dimethyl-1H-indene	$C_{11}H_{14}$	2.91
14	14.278	1,3-Dimethyl-1H-indene	$C_{11}H_{12}$	3.39
18	15.8	1-Tridecene	$C_{13}H_{26}$	7.41
19	16.21	2-Methyl-naphthalene	$C_{11}H_{10}$	10.95
21	19.851	1-Tetradecene	$C_{14}H_{28}$	8.42
22	20.172	Tetradecane	$C_{14}H_{30}$	3.84
23	20.515	2,7-Dimethyl naphthalene	$C_{12}H_{12}$	2.79
26	23.818	1-Pentadecene	$C_{15}H_{30}$	5.99
27	24.122	Pentadecane	$C_{15}H_{32}$	3.88
29	27.636	1-Hexadecene	$C_{16}H_{32}$	4.84
30	27.919	Hexadecane	$C_{16}H_{34}$	2.89
32	31.298	3-Heptadecene	$C_{17}H_{34}$	3.57
33	31.556	Heptadecane	$C_{17}H_{36}$	3.55
34	34.787	1-Octadecene	$C_{18}H_{36}$	2.42
35	35.032	Pentadecane	$C_{18}H_{37}$	2.4
37	38.123	1-Nonadecene	$C_{19}H_{38}$	1.76

PERFORMANCE TEST OF THE BLENDS IN A CI ENGINE

The experimental setup consists of a single cylinder Kirloskar CI engine (5 hp, 1500 rpm, 4 stroke, and 500 cc), which is mechanically loaded by means of a brake drum dynamometer. The performance of plastic oil blends in a CI engine was investigated and compared with pure diesel. Five blends of the oil were prepared. It includes B10 (10% oil and 90% diesel), B20 (20% oil and 80% diesel), B30 (30% oil and 70% diesel), B50 (50% oil and 50% diesel), and B80 (80% oil and 20% diesel). The engine is run using the above seven oils, and the results are compared.

The variation of fuel consumption with brake power is shown in Figure 4. With the increase in brake power the engine requires more energy and hence more fuel causing an increase in fuel consumption. It can be seen from the figure that with the increase in the concentration of the plastic oil in the blends, the fuel consumption goes on increasing; the reason is the plastic oil has a lesser calorific value than diesel (lower calorific value of plastic oil: 41.89 MJ/kg).

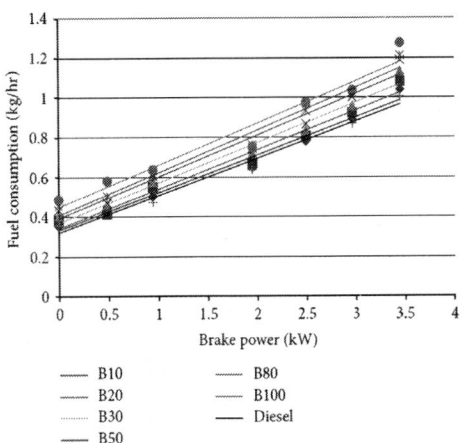

Figure 4: Fuel consumption versus brake power.

The variation of brake thermal efficiency with brake power is shown in Figure 5. Brake thermal efficiency is defined as the ratio of brake power output to the net input power. Brake thermal efficiency increases with brake power only up to a limit beyond which it drops due the incomplete combustion taking place. Here, with increase in the concentration of plastic oil in the blends, the efficiency decreases which is due to the higher fuel consumption, and the pure plastic oil gives the least efficiency.

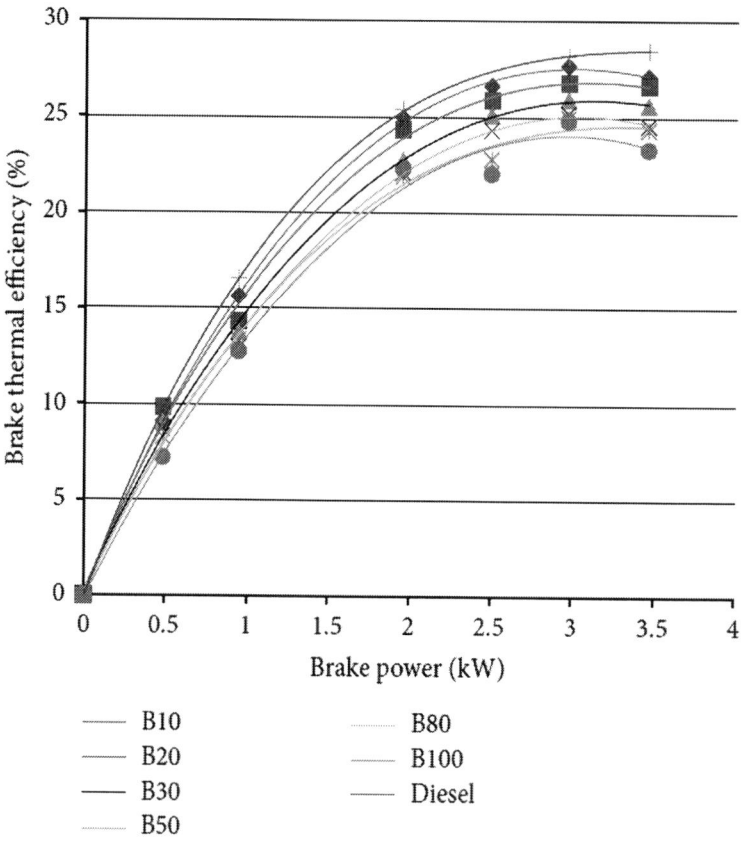

Figure 5: Brake thermal efficiency versus brake power.

The variation of indicated thermal efficiency with brake power is shown in Figure 6. Indicated thermal efficiency is defined as the ratio of indicated power output to the input power. Even though the frictional power increases with the blend %, the blends show less efficiency because of its higher fuel consumption. Interestingly it can be noticed in the figure that the higher blends and pure oil show a higher efficiency than its lower counterparts. This is because the increase in frictional power is less for higher blends as compared to lesser blends.

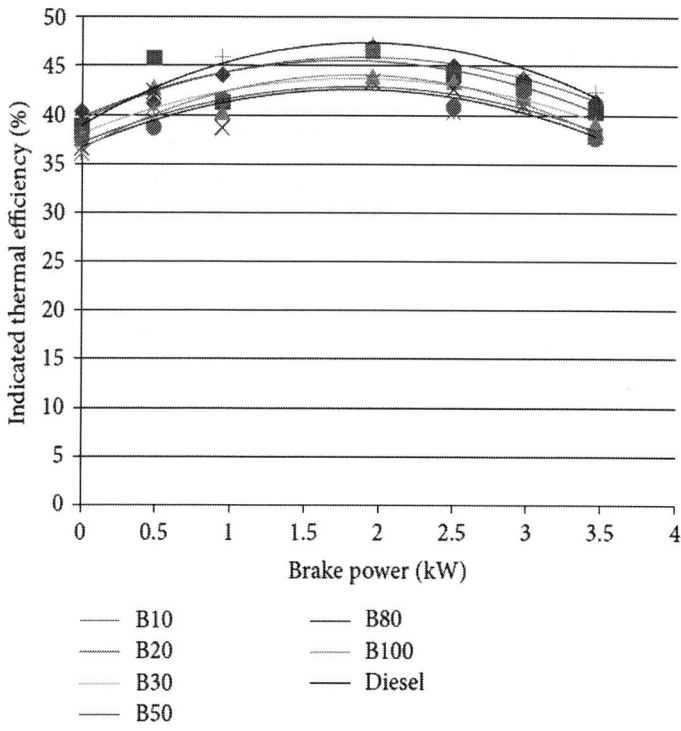

Figure 6: Indicated thermal efficiency versus brake power.

The variation of mechanical efficiency with brake power is shown in Figure 7. Mechanical efficiency is defined as the ratio of brake power to the indicated power. It can be seen that the mechanical efficiency decreases with the increase in blend concentration. This can be attributed to the increase in frictional power with the increase in blend %.

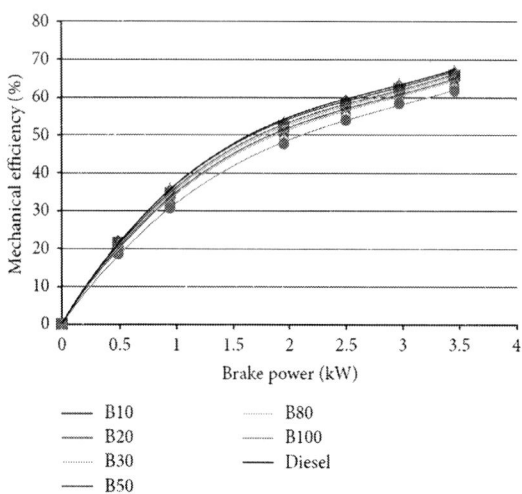

Figure 7: Mechanical efficiency versus brake power.

The variation of volumetric efficiency with brake power is shown in Figure 8. Since the test engine is a constant speed engine, volumetric efficiency remains the same irrespective of the fuel used.

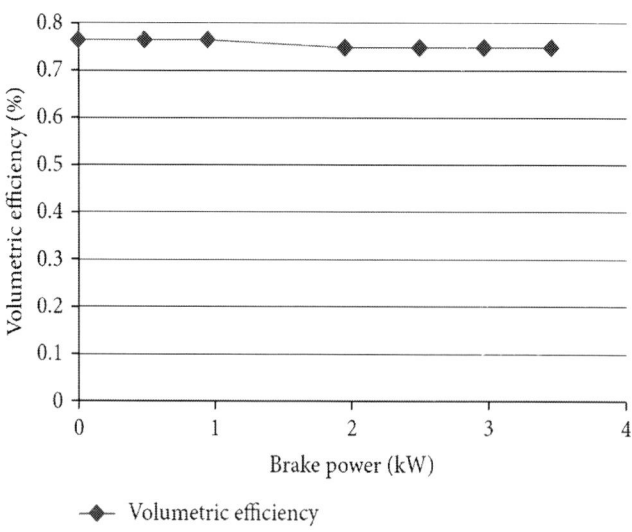

Figure 8: Volumetric efficiency versus brake power.

The variation of air-fuel ratio with brake power is shown in Figure 9. With increase in brake power, more will be the fuel consumption, and hence the air to fuel ratio will be decreasing. From the figure it can be seen that the air fuel ratio is going on decreasing with increase in the plastic oil concentration, which is due to the increasing fuel consumption with increasing blend %.

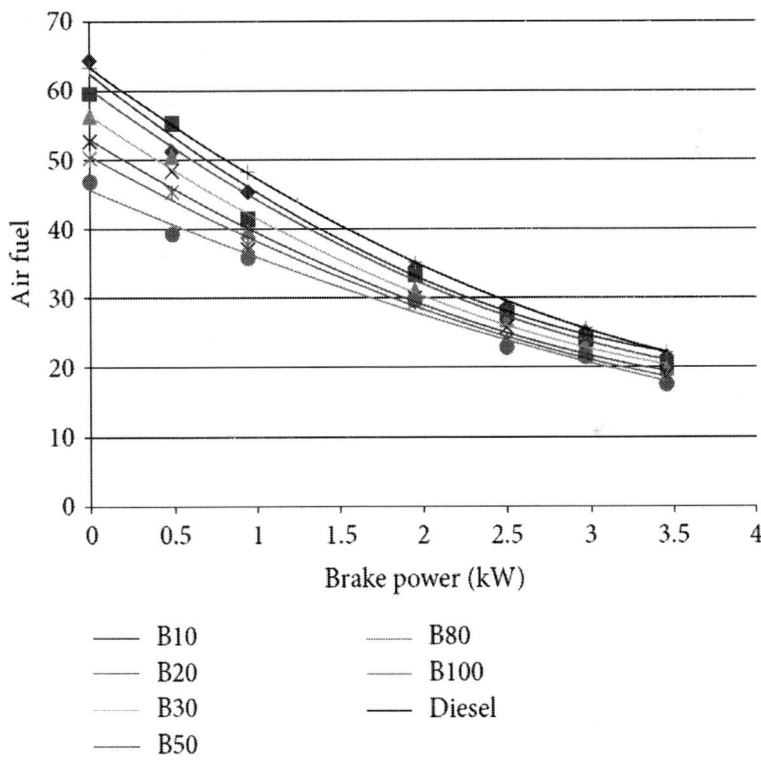

Figure 9: Air fuel ratio versus brake power.

The variation of brake specific fuel consumption with brake power is shown in Figure 10. Brake specific fuel consumption is defined as the amount of fuel required for producing unit brake power. The plastic oil and its blends have higher bsfc than diesel because of their higher fuel consumption arising from their lesser calorific value than diesel.

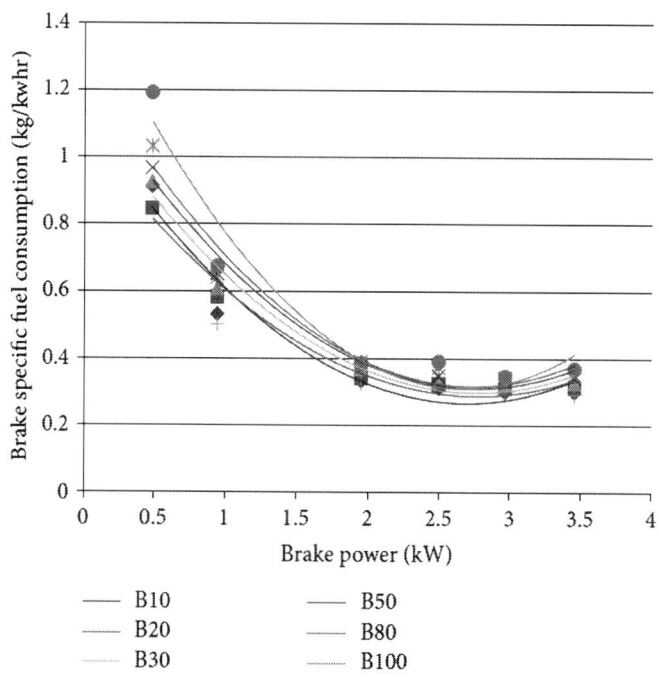

Figure 10: Brake specific fuel consumption versus brake power.

EMISSION CHARACTERISTICS

From the performance test analysis it can be seen that the test results of B20 (20% plastic oil and 80% diesel blend) showed close similarities with that of diesel. So for the analysis of the emission characteristics, the engine was run with B20 and diesel, and the emission was analysed using an exhaust gas analyser. Emissions analysed were CO emission and NO_x emission.

The variation of carbon monoxide content with brake power is compared for diesel and B20 as shown in Figure 11. It can be seen that the CO emission is lesser for B20 as compared to diesel.

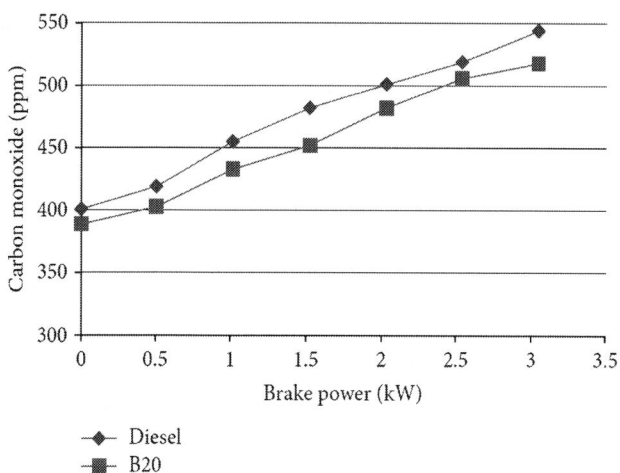

Figure 11: Carbon monoxide emission (ppm) versus brake power.

From the literature, it can be seen that with plastic oil there is an ignition delay of 2.5° crank angle. This ignition delay causes a steep rise in the peak pressure causing a high exhaust temperature. This is evident from Figure 12 which shows a comparison of the exhaust gas temperatures of B20 and diesel.

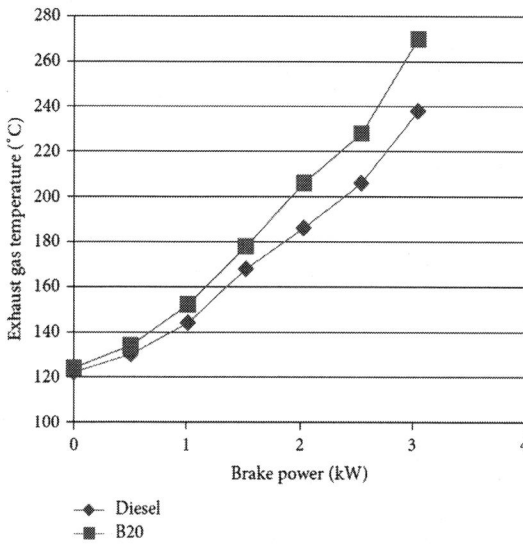

Figure 12: Exhaust gas temperature versus brake power.

The variation of NO_x emission with brake power is compared for diesel and B20 as shown in Figure 13. The exhaust gas temperature drives a direct relation with the NO_x emission. So it is clear that the B20 sample will give higher NO_x emission than with diesel.

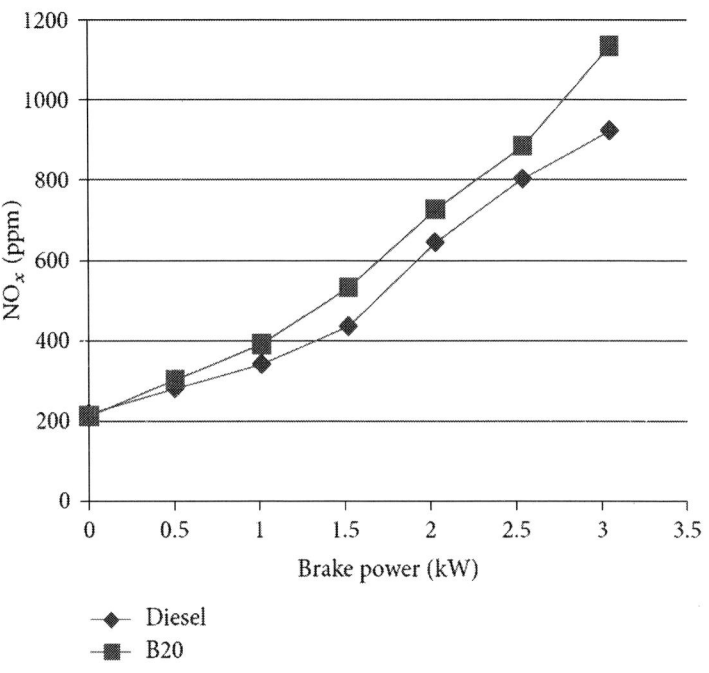

Figure 13: NO_x emission versus brake power.

CONCLUSIONS

- A petroleum based fuel has been produced from waste plastic (polythene).
- The optimum catalyst and reactions for catalytic pyrolysis of polythene have been found. Based on the yield and thermophysical properties, the combination of silica alumina and zeolite 1 (pore size ~4 Å) was selected as the optimum catalyst.

- The properties of the plastic oil and its chemical composition have been examined. The average chemical formula was found to be $C_{13.18}H_{23.56}$, and hence the performance analysis was done in a CI engine.
- In the performance analysis in engine, even though the plastic oil shows inferior results as compared to diesel, the lower blends percentage oils show results close with that of diesel (B10, B20, and B30). This makes it a strong competitor in the area of alternate fuels. Also the blend B20 has low CO emissions than for diesel. However, the NO_x emissions are higher for B20.
- 64.15% of the production cost is accounted for the cost of catalysts. If cheaper catalysts can be employed, the production cost can be decreased considerably.
- If the gaseous products and solid can be used, then the effective cost will come down even further.
- Rather than considering it just as an alternate fuel, the practical importance of this method in waste plastic management adds its value as an alternate fuel.

REFERENCES

1. A. K. Panda, R. K. Singh, and D. K. Mishra, "Thermolysis of waste plastics to liquid fuel: a suitable method for plastic waste management and manufacture of value added products: a world prospective,"Renewable and Sustainable Energy Reviews, vol. 14, no. 1, pp. 233–248, 2010.
2. http://cipet.gov.in/plastics_statics.html.
3. A. R. Songip, T. Masuda, H. Kuwahara, and K. Hashimoto, "Test to screen catalysts for reforming heavy oil from waste plastics," Applied Catalysis B, vol. 2, no. 2-3, pp. 153–164, 1993.
4. G. Manos, A. Garforth, and J. Dwyer, "Catalytic degradation of high-density polyethylene on an ultrastable-Y zeolite.

Nature of initial polymer reactions, pattern of formation of gas and liquid products, and temperature effects," Industrial and Engineering Chemistry Research, vol. 39, no. 5, pp. 1203–1208, 2000.

5. M. A. Uddin, K. Koizumi, K. Murata, and Y. Sakata, "Thermal and catalytic degradation of structurally different types of polyethylene into fuel oil," Polymer Degradation and Stability, vol. 56, no. 1, pp. 37–44, 1997.

6. M. R. Jan, J. Shah, and H. Gulab, "Catalytic degradation of Waste high-density polyethylene into fuel products using $BaCO_3$ as a catalyst," Fuel Processing Technology, vol. 91, no. 11, pp. 1428–1437, 2010. View at Publisher

7. N. S. Akpanudoh, K. Gobin, and G. Manos, "Catalytic degradation of plastic waste to liquid fuel over commercial cracking catalysts: effect of polymer to catalyst ratio/acidity content," Journal of Molecular Catalysis A, vol. 235, no. 1-2, pp. 67–73, 2005.

8. H. Gulab, M. R. Jan, J. Shah, and G. Manos, "Plastic catalytic pyrolysis to fuels as tertiary polymer recycling method: effect of process conditions," Journal of Environmental Science and Health, vol. 45, no. 7, pp. 908–915, 2010.

Chapter 2

Molecular Components-based Representation of Petroleum Fractions

Muhammad Imran Ahmad[a], Nan Zhang[b], and Megan Jobson[b]

[a]Department of Chemical Engineering, University of Engineering and Technology, 25000 Peshawar, Pakistan

[b]Centre for Process Integration, School of Chemical Engineering and Analytical Science, The University of Manchester, Sackville Street, Manchester M60 1QD, UK

ABSTRACT

Characterisation of petroleum fractions is the systematic analysis and representation of composition and properties of petroleum fractions. Characterisation methods play an important role in understanding of the physical and chemical behaviour of a petroleum fraction, its individual constituents, and are essential for modelling of refinery processes. In order to comply with the current and future product specifications for cleaner fuels, refineries are employing new processing technologies and more severe operating conditions in existing operating units. Consequently refinery process models are required to capture the chemistry of conversion processes employing characterisation of petroleum fractions at molecular level. In this paper a review of the conventional characterisation methods used for modelling of refinery processes is presented. The molecular type and homologous series (MTHS) matrix representation of petroleum fractions is discussed in detail. The previous work on MTHS matrix representation approach is limited to light petroleum fractions such as gasoline, and takes into account only hydrocarbon molecules. These shortcomings of MTHS matrix representation approach are addressed in this work through the development of a new strategy for estimation of composition and properties of petroleum fractions.

INTRODUCTION

A petroleum fraction is considered well-defined if the composition and structure of all its constituents is known (Riazi, 2005). However, a comprehensive analysis of a petroleum fraction is a very tedious task. For this reason refinery processes have been modelled conventionally using lumped models (Saine Aye, 2003). Some of the most important ways of expressing composition of petroleum fractions, some of which have been used extensively for conventional refinery process models, are (Riazi, 2005):

- PNA (paraffins, naphthenes, and aromatics);
- PONA (paraffins, olefins, naphthenes, and aromatics);
- PIONA (paraffins, iso paraffins, olefins, naphthenes, and aromatics);
- SARA (saturates, aromatics, resins, and asphaltenes);
- Elemental analysis (C, H, S, N, O).

Conventional Characterisation Methods

Three characterization methods have been widely used for modelling refinery processes. These are bulk property characterisation, compound class characterisation and average structural property characterisation methods. A brief introduction of these methods for characterisation of petroleum fractions is given here and detailed accounts can be found elsewhere (Zhang, 1999 and Saine Aye, 2003).

Bulk property characterisation methods are based on the classification of petroleum fractions in terms of boiling point cuts using bulk properties such as true boiling point curve, density, and API gravity, etc. Commercial simulation packages such as ASPENPLUS, HYSYS (AspenTech), and PROII (Simulation Sciences) make use of these methods to characterise streams in terms of pseudo-components, treating these pseudo-components as real components for simulation purposes.

Compound class characterisation methods are based on chromatographic separation techniques, taking into account the solubility characteristics of the components of a mixture. A typical example of compound class characterisation is the SARA (saturates, aromatics, resins, and asphaltenes) method. The SARA method, e.g. Speight (1998), is suitable for characterisation of heavy petroleum fractions, residues, and coal liquids (Riazi, 2005).

Average structural parameter characterisation methods classify petroleum mixtures on the basis of functional groups or structural parameters such as the number of aromatic rings, carbon/hydrogen ratio in an aromatic structure, weight percentage of carbon atoms in CH_2 groups, etc. Some of the examples of this type of characterisation methods are the Brown–Ladner method (Brown and Ladner, 1960), computer-assisted molecular construction (Oka et al., 1977), and functional group analysis by linear programming (Petrakis et al., 1983).

These methods broadly classify mixtures on the basis of existing compound classes or overall characteristics, such as bulk properties or average structural parameters, and do not provide information about the individual components. In this era of cleaner fuel specifications, to ensure more environmentally benign fuels, it is essential to obtain information of individual components of various compound classes existing in a petroleum fraction as the physical properties, vapour–liquid equilibrium characteristics, and reactivity of components change with molecular structure. Such detailed level of information has become fundamental to refinery process modelling due to stringent product specifications.

Chemical Analysis for Characterisation of Petroleum Fractions

Modern analytical instruments analyse the composition and properties of petroleum fractions at molecular level, i.e. providing composition and structural information of individual components of compound classes existing in a mixture. These analytical techniques provide the information expressed systematically in characterisation methods and can be classified into three different categories (Riazi, 2005):

- Chromatographic techniques;
- Separation by solvents;
- Spectroscopic and spectrometric methods.

Gas chromatography is a technique for analysis of petroleum fractions by separating components of a mixture based on volatility of compounds (Riazi, 2005). This technique is used for detailed analysis of the gasoline range, i.e. fractions boiling below 180 °C, but has limitations for fractions with high boiling point range due to low volatilities of components (Saine Aye, 2003). Teng et al. (1994) developed a method for detailed analysis of hydrocarbon components of gasoline using gas chromatography with mass spectrometry. Liquid chromatography is typically applied for separation of aromatic compounds according to the number of aromatic rings in heavy petroleum fractions.

The separation by solvents method is based on the solubility of components of a mixture in a particular solvent thereby separating soluble components from insoluble ones. This method is useful for the SARA analysis of heavy petroleum fractions and residues (Riazi, 2005).

The third category of chemical analysis of petroleum fractions is based on the identification of individual molecular groups present in a petroleum fraction using spectroscopic or spectrometric methods: In spectroscopic methods, molecules are excited by various sources and returned to their normal state, while in spectrometric methods different molecules present in a mixture are ionised and fragmented for composition analysis (Riazi, 2005). Nuclear magnetic resonance (NMR) spectroscopy is an example of spectroscopic analysis which, combined with elemental analysis, can measure the aromatic and aliphatic carbon, hydrogen distributions and concentrations of various structural groups (Zhang, 1999). Mass spectrometric methods analyse the composition of petroleum fractions based on the molecular weight and chemical structure of the components present in the mixture and provide the most detailed analysis (Riazi, 2005).

These analytical techniques provide information about the composition, structure, and properties of individual components of petroleum fractions and facilitate molecular components-based modelling of refinery processes.

Characterisation of Petroleum Fractions with Empirical Correlations

Characterisation methods using empirical correlations for the estimation of composition of petroleum fractions, in terms of PNA (paraffins, naphthenes, and aromatics) contents, using bulk physical properties such as molecular weight, normal boiling point, and specific gravity are extensively available in literature (e.g. Manafi et al., 1999, El-Hadi and Bezzina, 2005, Behrenbruch and Dedigama, 2007 and Choudhary and Meier, 2008). These correlation-based methods predict the PNA contents, and phase behavior of petroleum fluids reasonably accurately. However, these correlations are developed based on the properties of pure hydrocarbons, and may result into inaccuracies for mixtures containing hetero-atoms such sulphur and nitrogen containing compounds (Aladwani and Riazi, 2005).

Molecular Reconstruction of Composition of Petroleum Fractions

The current specifications for cleaner transportation fuels require representation of composition of petroleum fractions at molecular level for modelling and optimisation of refinery processes. Modern characterisation techniques that generate composition of petroleum fractions in terms of individual representative molecules, utilising only the readily available analytical data are known as molecular reconstruction techniques (Verstraete et al., 2010). Neurock (1992) and Neurock et al. (1994) developed stochastic reconstruction method to generate composition of petroleum fractions using distributions of molecular structural attributes through Monte-Carlo sampling. The stochastic reconstruction approach has been adopted and modified in recent efforts together with stochastic optimisation algorithms in order to minimise the difference between the calculated and experimental data (Hudebine and Verstraete, 2004, Van Geem et al., 2007 and Verstraete et al., 2010). The molecular reconstruction techniques, using stochastic reconstruction along with stochastic optimisation, may be employed to rebuild mixtures that resemble the petroleum fractions more closely than the approaches used previously. However, the computation time of molecular reconstruction techniques may be a drawback in applications involving optimisation of refinery processes. Another limitation of the molecular reconstruction techniques, discussed in this Section, is the exceeding number of molecules sampled to represent a petroleum fraction. For example, a set of 5000 molecules was obtained to represent an Arabian Light Vacuum residue fraction by Verstraete et al. (2010). Such exceeding number of representative molecules may become a shortcoming, in terms of predicting reaction kinetics, while modelling reactions of hydrocarbon conversion processes. A molecular characterisation approach that is numerically tractable, compared to the molecular reconstruction techniques discussed here, is presented in the next section.

MOLECULAR TYPE AND HOMOLOGOUS SERIES (MTHS) MATRIX REPRESENTATION OF PETROLEUM FRACTIONS

Peng (1999) proposed a matrix representation to incorporate the composition and structural information of individual components of mixtures available due to modern analytical techniques for chemical analysis of petroleum fractions. This characterisation approach is known as molecular type and homologous series (MTHS) matrix representation of petroleum fractions and originates from the structure-oriented lumping (SOL) approach, developed by Quann and Jaffe (1992), for describing the composition, reactions, and properties of complex hydrocarbon mixtures. The approach proposed by Quann and Jaffe (1992) employs a set of vectors with incremental features to represent the structure of individual molecules present in a mixture.

The molecular type and homologous series (MTHS) matrix characterisation approach represents the composition of a petroleum fraction in terms of homologous series and carbon number information. The columns of the matrix are the molecular types existing in the petroleum fraction, e.g. paraffins, aromatics, and naphthenes while the rows represent the carbon numbers, i.e. the molecular size of components. The size of molecular type and homologous series matrix, i.e. the different compound classes forming matrix columns and carbon number range determining the number of rows of the MTHS matrix may vary from one case to another depending on the petroleum fraction represented. Some of the compound classes, used to represent middle distillate diesel fractions in this work, are as follows:

nP, iP: normal and iso-paraffins;

1N, 2N, 3N: naphthenic compounds containing up to three rings, e.g., decalin ⬡⬡ belongs to the **2N** column;

A, 2A, 3A: one-ring to three-ring aromatic compounds: e.g. ⬡⬡⬡ and ⬡⬡⬡ are isomers that are both included in the same lump of the **3A** column;

AN, AAN, ANA, ANN: compounds containing both naphthenic and aromatic rings, e.g. the **AN** column contains ⬡⬡ and ⬡⬡~ ;

A_A, A_AA: biphenyl type compounds, e.g. ⬡-⬡ and ⬡-⬡-⬡;

A_N: cyclohexylbenzene, which is important as a hydrodesulphurisation product;

SI, SII, SIII, SIV, and SV: sulphur containing compounds:

SI: mercaptans, e.g. CH_3SH, C_2H_5SH;

SII: sulphides, disulphides, thiophenes and benzothiophenes, e.g. ⬡-S, CS_2, CH_3-S-C_2H_5 all belong to this column;

SIII: dibenzothiophenes not substituted at the 4 or 6 position, e.g. ⬡-S-⬡;

SIV: dibenzothiophenes substituted at the 4 or 6 position, e.g. ⬡-S-⬡ ;

SV: dibenzothiophenes substituted at both 4 and 6 positions, e.g. ⬡-S-⬡ ;

NI, NII: nitrogen containing compounds:

NI: basic nitrogen compounds, e.g. pyridine ⬡N and quinoline ⬡⬡N ;

NII: non-basic nitrogen compounds, e.g. indole ;

The elements of the matrix may represent the composition of a mixture on molar, weight or volume basis. Each matrix entry represents either a single compound or a lump of structural isomers. Fig. 1 shows the structure of molecular type and homologous series matrix. For example, the C_1, C_2, and C_3 rows under the nP column in Fig. 1 represent the composition of methane, ethane, and propane. The first row of the matrix i.e. C_0 consists of compounds without any carbon atoms. This row of the matrix captures the information of important inorganic molecules such as Hydrogen (C_0, nP), hydrogen sulphide (C_0, SI), and ammonia (C_0, NI). The sulphur compounds in middle distillates are classified into five groups: SI to SV. SI homologous series includes mercaptans and similar compounds while the other four groups account for alkyl benzothiophenes (SII), dibenzothiophenes and alkyl dibenzothiophenes without substituents at the 4- and 6-positions (SIII), alkyl dibenzothiophenes with only one of substituents at either the 4- or 6-position (SIV), and alkyl dibenzothiophenes with two of the alkyl substituents at the 4- and 6-positions.

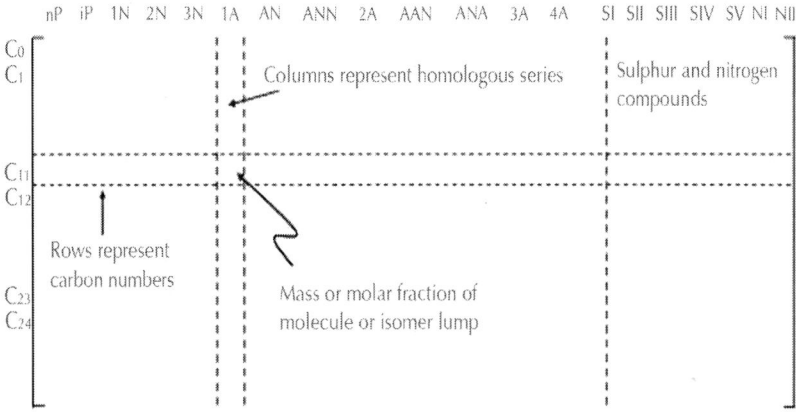

Figure. 1: Molecular type and homologous series matrix representation of petroleum fractions (Peng, 1999).

The matrix framework lumps all the structural isomers of a particular molecular size into one matrix entry. The reason behind this lumping simplification is that the current analytical techniques provide information of all structural isomers existing in a petroleum fraction for only lower carbon numbers, i.e. gasoline range fractions. This lumping of isomers can be a shortcoming of the MTHS matrix framework for applications in the modelling and optimisation of azeotropic distillation processes. On the other hand this strategy of lumping structural isomers as one matrix element reduces the complexity of the possible molecular model with around 10^{14} molecules to less than 10^3 matrix components (Peng, 1999).

The analytical techniques introduced in the Section 1.2 provide composition analysis for petroleum fractions up to C_{45} (Boduszynski, 1988). Fafet and Magne-Drisch (1995) analysed petroleum fractions in the middle distillate range using a combination of chromatographic and spectrometric techniques. The composition (wt %) obtained from this analysis is shown in Fig. 2.

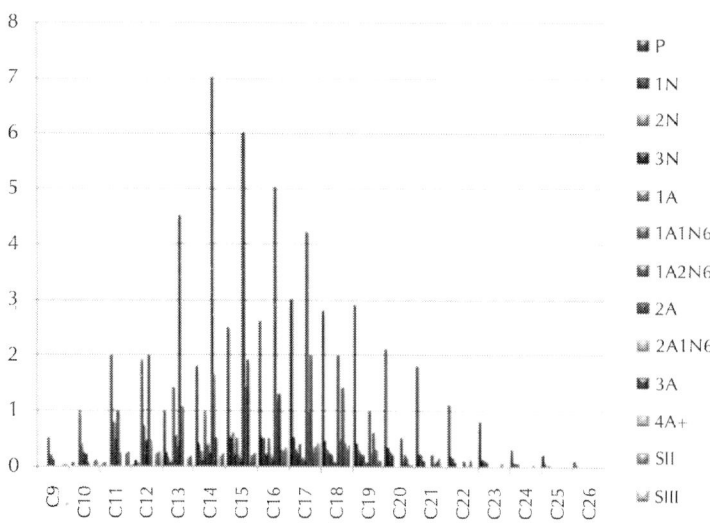

Figure. 2: Composition analysis of middle distillate petroleum fraction (Fafet and Magne-Drisch, 1995).

With this detailed level of analysis, composition can be directly represented in the molecular type and homologous series matrix framework.

Molecular Type and Homologous Series Matrix Generation

The analytical techniques introduced in Section 1.2 for the molecular analysis of petroleum fractions can be very expensive and time-consuming and are therefore not employed in refining operations (Saine Aye, 2003). In order to avoid experimental measurements for obtaining molecular composition of petroleum fractions, previous researchers have developed methods for generation of molecular composition matrices. The concept of molecular type and homologous series matrix generation, i.e. methods to determine the composition of individual matrix entries without comprehensive chemical analysis, relies on the assumption that properties of a mixture are resultant of the properties of individual components and thus depend on the mixture composition. This assumption implies that if the bulk properties of a mixture and the individual component properties are available then the composition of the mixture can be predicted using representative compound classes of the petroleum fraction under consideration.

Zhang (1999) proposed a method of generation of molecular type and homologous series matrix for composition of petroleum fractions using interpolation of available molecular composition matrices. This method is based on the principle that the molecular composition of a petroleum fraction can be obtained by treating this stream as a blend of several oil streams of known molecular compositions. Saine Aye and Zhang (2005) generated sample matrices, i.e. composition matrices representative of available chemical analysis information, of various refinery product streams to serve as a database for MTHS matrix generation using interpolation method. Gomez-Prado et al. (2008) proposed an alternative representation of matrix framework known as modified MTHS (mMTHS) matrix representation of petroleum fractions. In this modified MTHS matrix representation the molecular size of each component/cut is represented by its boiling point instead of carbon number. The composition matrix for a petroleum fraction is generated by minimising the difference between the calculated and measured bulk properties using characterisation parameters such as refractivity intercept and viscosity gravity constant.

The interpolation method for generation of molecular type and homologous series matrices relies on the availability of experimental data from chemical analysis of petroleum fractions. This dependence on experimental data limits the application of this method to only those petroleum fractions for which molecular level chemical analysis is available in literature.

Estimation of Physical Properties

The aim of molecular modelling of refinery processes is to predict more accurately the performance of hydrocarbon conversion processes and the properties of the resulting product streams. Therefore, methods for estimating physical properties of mixtures are needed for MTHS matrix framework. Zhang (1999) developed a strategy for predicting bulk properties of petroleum streams from the molecular information using molecular structure-property correlations for normal boiling point and density of matrix components. The individual component boiling points and densities are used to estimate the bulk density and boiling point curve of the petroleum mixture. The bulk density can be calculated as a weighted average using component densities assuming an ideal mixing rule for petroleum fractions (Jabr et al., 1992 and Riazi and Al-Sahhaf, 1995). The true-boiling point (TBP) curve of a mixture is a plot of the individual boiling points of pure hydrocarbon compounds in increasing order against the cumulative volume percentage (Nelson, 1958). Based on this definition the true-boiling point curve of a petroleum fraction can be developed by plotting the normal boiling temperatures of matrix elements against the corresponding cumulative volume composition in an ascending order. Once the bulk density and boiling point curve of the mixture is obtained desired physical properties, such as aniline point, cetane number, cloud point and pour point, are estimated with standard correlations available in the literature (Lee and Kesler, 1975, Twu, 1985,Riazi and Daubert, 1987 and Cookson et al., 1995).

Limitations of MTHS Matrix Characterisation Approach

The molecular type and homologous series matrix representation of petroleum fractions forms the basis of the molecular components-based modelling approach followed in this work for hydrotreating process design. However, the previous work on the MTHS matrix characterisation approach, presented in Sections2.1 and 2.2, has some limitations which are discussed in this section and addressed in this work.

The interpolation method (Zhang, 1999) for generation of MTHS matrices requires a database of sample matrices of the same range of boiling point and properties as the desired petroleum fraction. Such sample composition matrices are available in the literature for gasoline range but not for higher boiling petroleum fractions, for generation of molecular type and homologous series matrices by the interpolation method.

The strategy followed by Zhang (1999) and Saine Aye and Zhang (2005) for the estimation of properties of petroleum fractions using the matrix framework has a limited scope of application. The main reason for this shortcoming is that the molecular structure–property correlations developed by Zhang (1999) do not account for hydrocarbons with hetero-atoms such as sulphur and nitrogen compounds. The hydrocarbons with hetero-atoms have significant importance in fuel products especially in diesel product specifications, e.g. maximum sulphur content, due to environmental concerns regarding emissions from transport vehicles. The molecular structure–property correlations are available only for low boiling hydrocarbons fractions, i.e. gasoline range, and not for compound classes existing in middle distillates and heavy petroleum fractions, such as 4A, 1A3N, 2A1N, 2A2N, 3A1N and 4N.

PROPOSED IMPROVEMENTS TO MTHS MATRIX CHARACTERISATION APPROACH

This section addresses the limitations of previous work on generation of molecular type and homologous series matrices and estimation of physical properties of petroleum fractions. The strategy and methods developed in this section aim to extend the scope of molecular type and homologous series matrix representation to middle distillate and heavier petroleum fractions.

The approach proposed in this work for generation of MTHS matrices is similar to the approach developed by Gomez-Prado et al. (2008) wherein a composition matrix for a petroleum fraction is generated by minimising the difference between calculated and measured bulk properties. However, in this work the bulk properties of petroleum fractions are calculated using the individual component properties instead of using characterisation parameters as proposed by Gomez-Prado et al. (2008). Another notable difference between the two methods is the estimation of physical and thermodynamic properties of individual components of a petroleum fraction. This work employs group contribution methods (discussed briefly in Section 3.1) to estimate the physical and thermodynamic properties of individual components of petroleum fractions based on the molecular structure of compounds. The previous work, on MTHS matrix representation, employs molecular structure–property correlations developed by Zhang (1999) for estimation of properties of petroleum fractions. The application of group contribution methods for estimation of individual component properties is presented for the first in this work, for generation of MTHS matrices. In this way, physical and thermodynamic properties of hydrocarbons, as well as sulphur and nitrogen containing compounds may be estimated for improved predictions of properties of heavy petroleum fractions compared to the approach of Gomez-Prado et al. (2008).

New Strategy for Estimation of Properties

The strategy employed in this work for the estimation of bulk properties of petroleum fractions is to estimate the individual component properties using group contribution methods (Lydersen, 1955 and Joback and Reid, 1987). Group contribution methods are based on the principle that the properties of a compound are function of the atoms and structural groups combining to form the compound, and estimate the physical properties of a species by using the contributions that have been assigned to different atoms and atomic groups for each type of physical constant (Prausnitz et al., 2001). Once the individual component properties are estimated, bulk properties of the mixture may be estimated using established mixing rules. The bulk properties estimated this way are employed in the approach, presented in next section, for the generation of sample MTHS matrices.

New Approach for Generation of MTHS Matrix

As discussed in the Section 2.3, the molecular composition by interpolation method for generation of the molecular type and homologous series matrix is limited to low boiling petroleum fractions, e.g. gasoline, and not used for middle distillate and heavy petroleum fractions due to the insufficient molecular composition data. In this work the approach developed by Saine Aye and Zhang (2005) for generation of sample matrices is modified by accounting for the carbon number distribution, i.e. the distribution of amounts of individual components with different number of carbon atoms in molecules, in various compound classes existing in the middle distillate petroleum streams.

It may observed, based on chemical analysis of petroleum fractions (Amorelli et al., 1992, Teng et al., 1994, Mushrush et al., 1999, Georgina et al., 2001, Laredo et al., 2002 and Bacha et al., 2007) that various compound classes existing in petroleum mixtures exhibit characteristic distributions. For example, the carbon number distribution for middle distillate fractions based on the measurements of Bacha et al. (2007) is shown in Fig. 3.

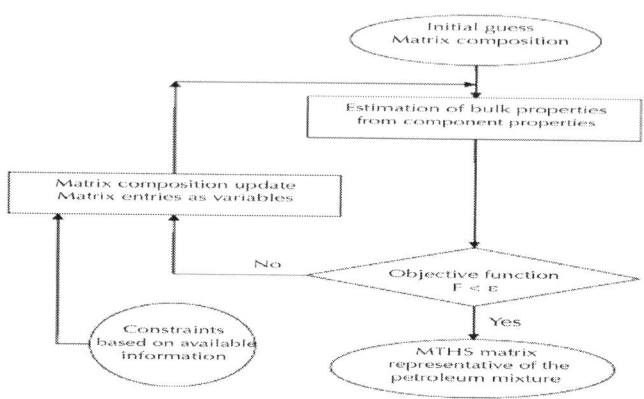

Figure. 3: Carbon number distribution for No. 2 diesel fuel (Bacha et al., 2007).

Sun (2004) identified and quantified the distribution of sulphur and nitrogen species in middle distillates. The carbon number distribution shown in Fig. 3 for the compound classes existing in diesel range petroleum streams is taken into account for modifying the approach developed by Saine Aye (2003) for generation of molecular type and homologous series matrices of diesel hydrotreating process streams. The mathematical formulation of the model proposed for generation of molecular type and homologous series matrix without the use of sample matrices is as follows.

The objective is to minimise the least squares difference of the measured and calculated properties (Zhang, 1999) of the middle distillate stream under consideration.

$$F = \sum \left(\frac{p^{measured} - p^{calculated}}{p^{measured}} \right)^2 \qquad (1)$$

The bulk properties of the stream are calculated using individual component properties and established mixing rules:

$$P^{calculated} = \sum_j \sum_i x_{i,j} \cdot p_{i,j} \qquad (2)$$

Where $x_{i,j}$ is the mole fraction and $p_{i,j}$ is the physical property of the matrix entry in the ith row and jth column of the molecular type and homologous series matrix. For some physical properties such as density and specific gravity, volume fraction is used in the above equation or alternately the following mixing rule is used:

$$P^{calculated} = \sum_j \sum_i \frac{w_{i,j}}{p_{i,j}} \qquad (3)$$

Where $w_{i,j}$ is the weight fraction of each component. The total fraction of a compound class existing in the stream is used to obtain a composition matrix representative of the mixture under consideration, using information available from chemical analysis.

$$LWB \leq C_j \leq UPB \qquad (4)$$

Where LWB and UBP are the lower and upper limits for the fraction of a homologous series existing in the stream, C_j is the composition of a homologous series i.e. the cumulative fraction of components of a column of the matrix given by Eq. (5):

$$C_j = \sum_i x_{i,j} \qquad (5)$$

The characteristic carbon number distribution of petroleum fractions is incorporated in this modified approach to limit the search space to realistic solutions resulting in improved representation of petroleum fractions in matrix framework. The schematic of the proposed approach for generation of molecular type and homologous series matrix for a given petroleum fraction is shown in Fig. 4.

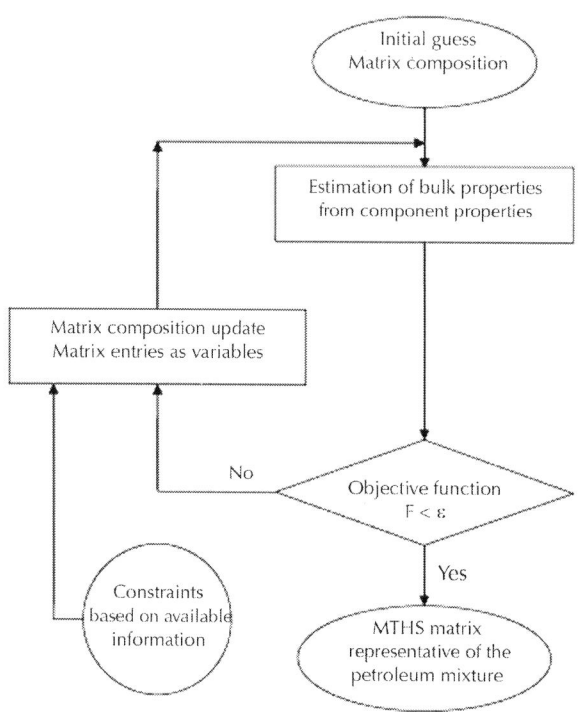

Figure. 4: Schematic of proposed approach for generation of molecular type and homologous series matrix.

In order to generate a composition matrix for a given petroleum fraction the proposed approach begins with an initial guess of composition. In this work, composition matrices based on experimental studies and from available literature have been employed in initialising the procedure for generation of MTHS matrices. However, a suitable initialisation methodology may improve the performance of the optimisation algorithm due to the multiplicity of solutions inherent to nature of such problems. The bulk properties of the composition matrix are calculated using component properties, i.e. properties of individual matrix entries, with established mixing rules (Jabr et al., 1992 and Riazi and Al-Sahhaf, 1995). The individual component properties are calculated using group contribution methods as discussed in Section 3.1. The objective function, i.e. the least squares difference of the measured and calculated properties given by Eq. (1), is minimised by varying the fractions of individual matrix entries which serve as the degrees of freedom. This proposed approach has been implemented using What'sBest!® 7.0 spreadsheet solver.

ILLUSTRATIVE EXAMPLES ON MTHS MATRIX GENERATION

In this section two examples are presented to illustrate the new approach, for generating molecular type and homologous series matrices, developed in this work. In the first example, a molecular composition matrix is generated for a synthetic crude distillate fraction from Canadian oil sands (Kimbara et al., 1996). The information available for the crude distillate fraction regarding the physical properties and composition are shown in Table 1.

Table 1: Physical properties and composition analysis of synthetic crude distillate (Kimbara et al., 1996)

Properties of synthetic crude distillate	
Density g/cm^3 (15 °C)	0.87
ASTM D86 distillation (wt% distilled) (°C)	
Initial boiling point	147
10%	196
30%	239
50%	265
90%	312
Final boiling point	354
Compositional mass spectrometric analysis (Mass %)	
Paraffins	20.1
Total cycloparaffins	40.4
Monocycloparaffins	21.7
Dicycloparaffins	12.3
Tricycloparaffins	6.4

Total aromatics	39.4
Alkylbenzenes	15.3
Benzocycloparaffins	11.7
Benzodicycloparaffins	4.4
Naphthalenes	5.1
Acenaphthenes	1.6
Fluorenes	0.5
Triaromatics	0.8

The molecular type and homologous series matrix generated using the previous approach of Zhang (1999) is shown in Table 2. The matrix entries shown in Table 2 are in weight %; * represents matrix components that do not exist. For example, matrix components in the column for di-aromatic compounds cannot exist for carbon numbers less than 10.

Table 2: Matrix composition of synthetic diesel fraction (Zhang, 1999)

	P	1N	2N	3N	1A	1A1N	1A2N	2A	2A1N	3A
C8	0	0.59	*	*	0.92	*	*	*	*	*
C9	0	0.7	0	*	1.84	0	*	*	*	*
C10	0	1.17	12.33	*	3.7	11.73	*	5.12	*	*
C11	0	2.83	0	*	3.51	0	*	0	*	*
C12	0	4.7	0	*	2.58	0	*	0	*	*
C13	0	5.87	0	0	1.84	0	0	0	0	*
C14	20.15	5.87	0	6.32	0.92	0	4.41	0	2.1	0.8

A comparison of these calculated properties, from the matrix generated using the method of Zhang (1999), with measured properties is shown in Table 3. The two boiling temperatures T10 and T90 are chosen for comparison as these values have been used in numerous correlations available in literature for estimation of other bulk properties (Cookson et al., 1995). It can be seen from Table 3 that there is large discrepancy in the true boiling point curve of the mixture estimated using the composition matrix generated with the approach of Zhang (1999). The reason behind this discrepancy is that in the previous work for generation of molecular type and homologous series matrix the characteristic carbon number range of petroleum fractions and the distribution of various compound classes is not considered.

Table 3: Comparison of properties calculated using the previous approach (Zhang, 1999)

Property	Measured	Calculated	Absolute Error
Density (g/cm^3)	0.87	0.87	0.0
T10 (°C)	196	194.2	1.8
T90 (°C)	312	282.5	29.5

The carbon number distribution employed in this work is based on the measurements of Bacha et al. (2007) as already shown in Fig. 3. The MTHS matrix obtained for the synthetic crude distillate using the approach developed in this work is shown in Table 4. The computation time required to obtain this composition matrix using the proposed approach is 1 min, and 10 CPU seconds (Pentium® D CPU 3.00 GHz processor and 1 GB RAM) with 555 function evaluations.

It can be seen from Table 4 that the molecular type and homologous series matrix generated to represent the composition of the petroleum fraction under consideration consists of the carbon number range characteristic of diesel streams. The carbon number distribution of compound classes existing in the mixture under consideration is shown in Fig. 5. A comparison of the distribution of the composition matrix generated using the proposed methodology with the distribution of the experimental findings of Fafet and Magne-Drisch (1995) is shown in Fig. 6. The cumulative weight percentages, of all the compound classes existing in a petroleum fraction, are plotted against the corresponding carbon numbers in Fig. 6. Fig. 6shows that the cumulative weight percentages of the composition matrix generated for the synthetic crude distillate under consideration are different from the cumulative weight percentages of the petroleum fraction analysed by Fafet and Magne-Drisch (1995).

Table 4: MTHS matrix (wt %) generated for synthetic crude distillate using the new approach for matrix generation

	P	1N	2N	3N	1A	1A1N	1A2N	2A	2A1N	3A
C8	0.509	0.473	*	*	0.351	*	*	*	*	*
C9	0.018	2.723	*	*	1.699	0.155	*	*	*	*
C10	1.600	1.977	0.318	*	0.749	0.331	*	0.010	*	*
C11	1.562	1.450	0.457	*	1.077	0.575	*	0.015	*	*
C12	2.088	1.938	0.611	*	1.440	0.819	*	0.020	*	*
C13	2.581	2.395	0.756	*	1.779	1.250	*	1.635	*	*
C14	3.209	2.978	5.903	1.942	2.212	2.132	2.050	2.210	1.670	0.304
C15	2.663	2.483	1.013	1.307	1.908	1.621	0.551	0.854	0.450	0.113
C16	2.281	2.145	0.805	1.264	1.575	1.297	0.537	0.798	0.233	0.112
C17	1.609	1.504	0.701	0.918	1.229	0.476	0.359	0.840	0.075	0.103
C18	1.249	1.075	0.855	0.642	0.854	0.404	0.337	0.312	0.071	0.095
C19	0.509	0.558	0.695	0.201	0.351	0.298	0.341	0.025	0.064	0.072
C20	0.221	0.000	0.185	0.125	0.077	0.290	0.226	0.015	0.051	0.000

The cumulative weight percentages are different in magnitudes, as expected, since the two petroleum fractions are from different crude sources. However, since both the petroleum fractions are middle distillates, the overall trends in composition distribution would be similar as can be seen from Fig. 6.

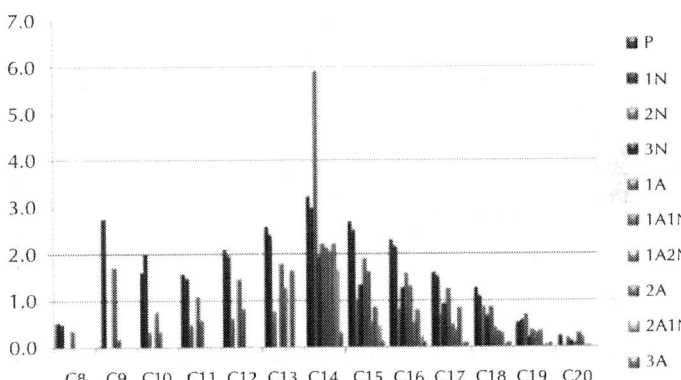

Figure. 5: Composition in wt% against carbon number, for synthetic crude distillate using the matrix generated with new approach for matrix generation.

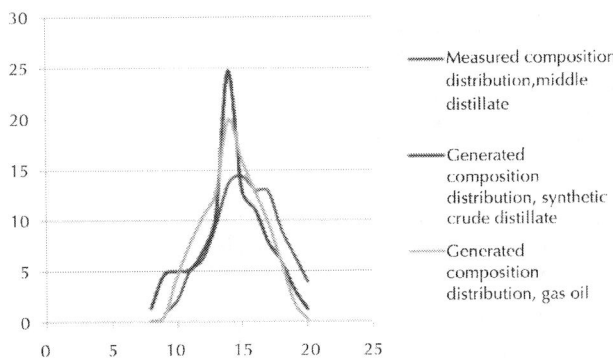

Figure. 6: Comparison of distribution of composition matrices for synthetic crude distillate and straight-run gas oil, with respect to carbon number, with the experimental findings of Fafet and Magne-Drisch (1995).

A comparison of some of the calculated bulk properties of the mixture with the measured bulk properties is shown in Table 5. It can be seen from Table 5 that the calculated bulk properties are consistently in good agreement with the measured properties. A comparison of the calculated amounts of paraffins, naphthenes, and aromatic (PNA) families with measured amounts indicates a very good agreement as shown in Table 6.

Table 5: Comparison of calculated bulk properties with measured bulk properties of Kimbara et al. (1996)

Property	Measured	Calculated	Absolute Error	Absolute % Error
Density (g/cm³)	0.87	0.861	0.009	1.062
T10 (°C)	196	192.694	3.306	1.687
T90 (°C)	312	309.310	2.690	0.862
M.wt. (g/gmol)	190.9	187.410	3.490	1.828

Table 6: Comparison of calculated PNA content with measured amounts from Kimbara et al. (1996)

	Measured (wt %)	Calculated (wt %)	Absolute % Error
Paraffins	20.1	20.09	0.05
Naphthenes	40.4	40.39	0.02
Aromatics	39.4	39.5	0.25

In the second example, a composition matrix is generated for a straight run gas oil (SRGO) stream that may serve as feedstock to diesel hydrotreating process (Sun, 2004). The properties of the feedstock are shown in Table 7.

Table 7: Properties of a straight run gas oil stream (Sun, 2004)

Feed	SRGO
Density (kg/m³)	830
Distillation (ASTM D86) (°C)	
Initial boiling point	155
0.1	230
0.3	260
0.5	275
0.7	300
0.9	325
Final boiling point	350
Total Sulphur (wt %)	0.3

The composition matrix generated using the proposed approach is shown in Table 8.

Table 8: MTHS matrix (wt %) generated for a straight run gas oil stream using the proposed approach for matrix generation

	P	N	NN	NNN	A	AN	ANN	AA	ANA	AAA	NI	NII	SI	SII	SIII	SIV	SV
C8	0.06	0	.	.	0.08	.	.	.	.	.	0.001	0	0.022	0	.	.	.
C9	0.11	0.09	.	.	0.21	0.08	.	.	.	.	0.007	0	0.037	0.002	.	.	.
C10	1.84	0.63	0.38	.	0.81	0.171	.	0.46	.	.	0.019	0	0.062	0.066	.	.	.
C11	2.64	1.42	1.24	.	1.17	0.25	.	0.72	.	.	0	0.004	0.037	0.126	.	.	.
C12	3.53	1.9	1.66	.	1.6	0.329	.	0.97	.	.	0	0.015	0.073	0.178	0.06	.	.
C13	4.36	2.4	2.1	.	1.93	0.406	.	1.2	.	.	0	0.018	0.069	0.099	0.146	0.001	.
C14	5.42	2.9	2.6	2.85	2.4	0.505	0.369	1.5	0.68	0.28	0	0.023	0.052	0.108	0.228	0.002	0.041
C15	4.36	2.4	2.1	2.29	1.93	0.406	0.49	1.2	0.54	0.22	0	0	0.051	0.048	0	0	0
C16	3.53	1.9	1.66	1.85	1.6	0.329	0.4	0.97	0.44	0.18	0	0	0.058	0	0	0	0
C17	2.64	1.5	1.3	1.38	1.17	0.246	0.301	0.72	0.33	0.13	0	0	0.056	0	0	0	0
C18	1.84	0.99	0	0.96	0.81	0.171	0.21	0.51	0.23	0.09	0	0	0	0	0	0	0
C19	0.86	0	0	0.13	0.1	0.09	0	0.63	0	0	0	0	0	0	0	0	0
C20	0	0	0	0.01	0.09	0	0	0.188	0	0	0	0	0	0	0	0	0

A comparison of the calculated amounts of paraffins, naphthenes, and aromatic (PNA) families with measured amounts indicates a good agreement as shown in Table 9.

Table 9: Comparison of calculated PNA content with measured amounts from (Sun, 2004)

	Measured (wt %)	Calculated (wt %)	Absolute % Error
Paraffins	30.95	31.19	0.78
Naphthenes	37.44	38.64	3.19
Aromatics	31.61	32.55	2.98

The carbon number distribution of compound classes existing in the mixture under consideration is shown in Fig. 7. A comparison of the distribution of the composition matrix generated with the distribution of the experimental findings of Fafet and Magne-Drisch (1995) is shown in Fig. 6. The trends in composition distribution for the second example are also similar to the trends observed by Fafet and Magne-Drisch (1995).

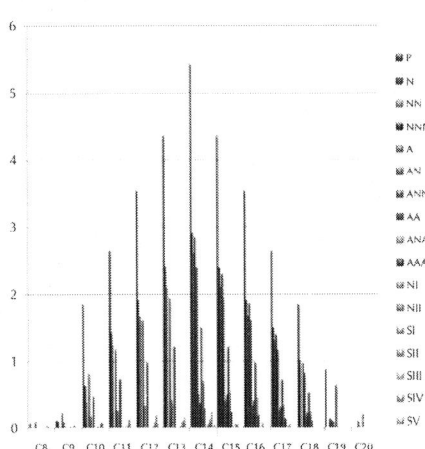

Figure. 7: Composition in wt% against carbon number, for straight run gas oil using the matrix generated with proposed approach for matrix generation.

The results of the two examples on matrix generation indicate that the proposed approach for matrix generation can provide a reasonable approximation for the composition, using a given distribution of compound classes existing in a petroleum fraction, in the absence of detailed composition analysis.

CONCLUSIONS

The conventional characterisation methods that are widely used in refinery process modelling do not provide detailed information at the molecular level. Such information is needed for modelling of refinery processes satisfying cleaner fuel specifications. The molecular type and homologous series matrix characterisation approach developed by Peng (1999) is a systematic representation of detailed molecular information that can be obtained using modern analytical techniques. In order to avoid expensive and time consuming chemical analysis of petroleum fractions previous researchers (Zhang, 1999, Saine Aye, 2003 and Saine Aye and Zhang, 2005) developed methods for generation of molecular type and homologous series matrices by minimising the difference between calculated and measured bulk properties of streams. The previous work on generation of molecular type and homologous series matrix representation is limited to low boiling petroleum fractions i.e. gasoline range. Another shortcoming of previous work is that the methods developed for estimation of physical properties of individual components of a mixture are limited to hydrocarbon compounds only.

A new strategy is developed in this work for estimation of bulk properties of petroleum fractions from the molecular composition matrix using group contribution methods to estimate properties of individual components, of not only hydrocarbon components but also hydrocarbons with hetero-atoms. Given these properties of individual components allows one to estimate the bulk properties of the mixture using mixing rules and correlations available in the literature. It has been shown that the new approach for matrix generation developed in this work can provide molecular composition information in the absence of detailed chemical analysis. The methods developed in this work extend the scope of molecular type and homologous series matrix characterisation approach to middle distillate and heavy petroleum fractions and thus facilitate molecular components-based modelling of heavy hydrocarbon conversion processes such as diesel hydrotreating process.

ACKNOWLEDGMENTS

Ahmad, M.I. would like to thank Prof. Robin Smith and Dr. Konstantinos Theodoropoulos for being a source of inspiration during various stages of Ph.D. study.

REFERENCES

1. Aladwani, H.A., Riazi, M.R., 2005. Some guidelines for choosing a characterisation method for petroleum fractions in process msimulators. Chemical Engineering Research and Design 83 (A2), 160–166.
2. Amorelli, A., Amos, Y.D., Halsig, C.P., Krosman, J.J., Joker, R.J., De Wind, M., Vrieling, J., 1992. Estimate feedstock processability, Hydrocarbon Processing, June, p. 93.
3. Bacha, J., Freel, J., Gibbs, A., Gibbs, L., Hemighaus, G., Hoekman, K., Horn, J., Ingham, M., Jossens, L., Kohler, D., Lesnini, D., McGeehan, J., Nikanjam, M., Olsen, E., Organ, R., Scott, B., Sztenderowicz, M., Tiedmann, A., Walker, C., Lind, J., Jones, J., Scott, D., Mills, J., 2007, Diesel Fuels Technical Review, Chevron Corporation.
4. Behrenbruch, P., Dedigama, T., 2007. Classification and characterisation of crude oils based on distillation properties. Journal of Petroleum Science and Engineering 57, 166–180.
5. Boduszynski, M.M., 1988. Composition of heavy petroleums. 2: Molecular characterisation. Energy & Fuels 2, 597–613.

6. Brown, J.K., Ladner, W.R., 1960. A study of the hydrogen distribution in coal-like materials by high-resolution nuclear magnetic resonance spectroscopy. II: A comparison with Infrared measurement and the conversion to carbon structure. Fuel 39, 87–96.
7. Choudhary, T.V., Meier, P.F., 2008. Characterisation of heavy petroleum fractions. Fuel Processing Technology 89, 697–703.
8. Cookson, D., Lliopoulos, P., Smith, B., 1995. Composition-property relations for jet and diesel fuels of variable boiling range. Fuel 74, 70–78.
9. El-Hadi, D., Bezzina, M., 2005. Improved empirical correlation for petroleum fraction composition quantitative prediction. Fuel 84, 611–617.
10. Fafet, A., Magne-Drisch, J., 1995. Quantitative-analysis of middle distillats by GC/MS coupling: application to hydrotreatment process mechanisms. Revue De L' Institut Francais Du Petrole 50, 404.
11. Georgina, C., Laredo, S., Luis Cano, D., Jesus Castillo, J.M., 2001. Inhibition effects of nitrogen compounds on the hydrodesulphurisation of dibenzothiophene. Applied Catalysis A: General 207, 103.
12. Gomez-Prado, J., Zhang, N., Theodoropoulos, C., 2008. Characterisation of heavy petroleum fractions using modified molecular-type homologous series (MTHS) representation. Energy 33, 974–987.
13. Hudebine, D., Verstraete, J.J., 2004. Molecular reconstruction of LCO gas oils from overall petroleum analyses. Chemical Engineering Science 59, 4755–4763.
14. Jabr, N., Alatiqi, I.M., Fahim, M.A., 1992. An improved characterisation method for petroleum fractions. The Canadian Journal of Chemical Engineering 70, 765–773.
15. Joback, K.G., Reid, R.C., 1987. Estimation of pure-component properties from group-contributions. Chemical Engineering Communications 57, 233–243.

16. Kimbara, N., Charland, J.P., Wilson, M.F., 1996. Hydrogenation of aromatics in synthetic crude distillates catalysed by platinum supported in molecular sieves. Industrial & Engineering Chemistry Research 35, 3874–3883.
17. Laredo, G.C., Leyva, S., Alvarez, R., Mares, M.T., Castillo, J., Cano, J.L., 2002. Nitrogen compounds characterisation in atmospheric gas oil and light cycle oil from a blend of Mexican crudes. Fuel 81, 1341–1350.
18. Lee, B.I., Kesler, M.G., 1975. A generalised thermodynamic correlation based on three parameter corresponding states. AIChE Journal 21, 510.
19. Lydersen, A.L., 1955. Estimation of Critical Properties of Organic Compounds by the Method of Group Contributions. Univ Coll Exp Stn Rept, Madison, WI.
20. Manafi, H., Mansoori, G.A., Ghotbi, S., 1999. Phase behavior prediction of petroleum fluids with minimum characterisation data. Journal of Petroleum Science and Engineering 22, 67–93.
21. Mushrush, G.W., Beal, E.J., Hardy, D.R., Hughes, J.M., 1999. Nitrogen compound distribution in middle distillate fuels derived from petroleum, oil shale, and tar sand sources. Fuel Process Technology 61, 197.
22. Nelson, W.L., 1958. Petroleum Refinery Engineering. McGraw-Hill, New York.
23. Neurock, M., 1992. A computational chemical reaction engineering analysis of complex heavy hydrocarbon reaction systems, Ph.D. Thesis, University of Delaware.
24. Neurock, M., Nigam, A., Trauth, D.M., Klein, M.T., 1994. Molecular representation of complex hydrocarbon feedstocks through efficient characterisation and stochastic algorithms. Chemical Engineering Science 49 (25), 4153–4177.
25. Oka, M., Chang, H.C., Gavales, G.R., 1977. Computer-assisted molecular structure construction for coal-derived Compounds. Fuel 56, 3–8.

26. Peng, B., 1999, Molecular modelling of refinery processes. PhD Thesis. UMIST, Manchester, UK.
27. Petrakis, L., Allen, D., Gavalas, G.R., Gates, B.C., 1983. Functionalgroup analysis of synthetic fuels. Analytical Chemistry 55, 1557.
28. Prausnitz, J.M., Poling, B.E., O'Connell, J.P., 2001. The Properties of Gases and Liquids, fifth ed. McGraw-Hill, New York.
29. Quann, R.J., Jaffe, S.B., 1992. Structure-oriented lumping: Describing the chemistry of complex hydrocarbon mixtures. Industrial & Engineering Chemistry Research 31, 2483–2497.
30. Riazi, M.R., 2005. Characterisation and Properties of Petroleum Fractions. ASTM International, Philadelphia, PA.
31. Riazi, M.R., Al-Sahhaf, T.A., 1995. Physical properties of n-alkanes and n-alkylhydrocarbons: application to petroleum mixtures. Industrial & Engineering Chemistry Research 34, 4145–4148. Riazi, M.R., Daubert, T.E., 1987. Characterisation parameters for petroleum fractions. Industrial & Engineering Chemistry Research 26, 755–759.
32. Saine Aye, M., 2003. Molecular modelling for cleaner fuel production. Ph.D. Thesis. Department of Process integration, UMIST, Manchester, UK.
33. Saine Aye, M., Zhang, N., 2005. A novel methodology in transforming bulk properties of refining streams into Journal Identification = CHERD Article Identification = 566 Date: March 23, 2011 Time: 1:33 pm 420 chemical engineering research and design 8 9 (2011) 410–420 molecular information. Chemical Engineering Science 60, 6702–6717.
34. Speight, J.G., 1998. The Chemistry and Technology of Petroleum, third ed. Marcel Dekker, New York.
35. Sun, J., 2004. Molecular modelling and integration analysis of hydroprocessors. Ph.D. Thesis. University of Manchester, UK.
36. Teng, S.T., Williams, A.D., Urdal, K., 1994. Detailed hydrocarbon analysis of gasoline by GC-MS (SI-PIONA).

Journal of High Resolution Chromatography, 17.

37. Twu, C.H., 1985. Internally consistent correlation for predicting liquid viscosities of petroleum fractions. Industrial & Engineering Chemistry Process Design and Development 24, 1287–1293.

38. Van Geem, K.M., Hudebine, D., Reyniers, M.F., Wahl, F., Verstraete, J.J., Marin, G.B., 2007. Molecular reconstruction of naphtha steam cracking feedstocks based on commercial indices. Computers and Chemical Engineering 31, 1020–1034.

39. Verstraete, J.J., Schnongs, Ph., Dulot, H., Hudebine, D., 2010. Molecular reconstruction of heavy petroleum residue fractions. Chemical Engineering Science 65, 304–312.

40. Zhang, Y., 1999. Molecular modelling of petroleum processes. Ph.D. Thesis. UMIST, Manchester, UK.

Chapter 3

Molecular Reconstruction of LCO Gasoils from Overall Petroleum Analyses

Damien Hudebine and Jan J. Verstraete

Development Division, IFP-Lyon, BP3, 69390 Vernaison, France

ABSTRACT

Reconstruction of petroleum fractions comprises the group of methods that allow to create mixtures of molecules from partial analytical data. In this article, a two-step reconstruction algorithm will be presented. The first step, called "stochastic reconstruction" step, assumes that oil mixtures can be described by distributions of structural blocks. The choice of the blocks and distributions is based on expert knowledge. The transformation from a set of distributions into a mixture of molecules is obtained by Monte-Carlo sampling, while a simulated annealing procedure adjusts the parametric distributions. The second step, termed "reconstruction by entropy maximization", improves the representativeness of the set of constructed molecules by adjusting their molar fractions. The estimation of these molar fractions is carried out by maximizing an information entropy criterion under linear constraints. The proposed two-step reconstruction algorithm allows to rebuild mixtures that resemble the petroleum fractions more closely than the approaches used previously. The technique is validated by comparing the properties of the rebuilt mixtures to analytical data that have not been used for the molecular reconstruction.

INTRODUCTION

In petroleum refining, both the accurate design and optimization of hydrocarbon conversion processes require the development of rigorous and reliable kinetic models. To establish an extensive kinetic model, the composition of the feedstock should be known with great detail. For the lightest products (gas, liquefied gas, gasoline), this characterization can be obtained by gas chromatography. For heavier cuts, however, no analytical technique is yet powerful enough to detect and quantify the thousands of different compounds that compose an oil fraction. Consequently, it is necessary to numerically "reconstruct" a mixture from partial analytical data, model hypotheses and expert knowledge.

Allen and Liguras (1991) proposed to select a set of predefined molecules and to modify their molar fractions in order to obtain a mixture whose properties are close to the desired analytical data. The analyses used are gas chromatography, ^{13}C NMR and ^1H NMR, and allow to set up 190 linear equality constraints as a function of the molar fractions. The problem solution is obtained by minimizing a specific criterion.

Quann and Jaffe (1996) suggested a similar method but the molecules are replaced by vectors of structural blocks, called "structure oriented lumping" (SOL) vectors. The reconstruction of the mixture of SOL vectors from analytical data has not been detailed by the authors for commercial reasons.

Zhang (1999) characterized petroleum cuts by a matrix of pseudo-compounds classified by chemical family and carbon number. Such a matrix can be obtained by coupling gas chromatography with mass spectrometry (GC/MS) but these analyses are long and complex. Another method consists of collecting a library of matrices associated with their overall mixture properties and calculating a new matrix by interpolation.

Others authors chose to define the petroleum fractions by an average model molecule in 2D (Hirsch and Altgelt, 1970 and Speight, 1970) or 3D (Faulon et al., 1990). The analyses used to build these average model molecules generally need to provide information on chemical functions, e.g. via infrared and NMR spectrometry.

Finally, Neurock et al. (Neurock, 1992 and Neurock et al., 1994) have developed a method called "stochastic reconstruction" (SR). The main hypothesis of this approach resides in the characterization of petroleum cuts by a set of distributions of molecular structural attributes. These distributions are sampled by a Monte-Carlo method in order to obtain a mixture of molecules. The method thus provides both a set of pseudo-compounds (molecule mixture) and a statistical average description (distributions). When coupled to an optimization loop on the distribution parameters, the method has been proven able to yield mixtures that closely mimic the original properties of heavy asphaltene feedstocks (Trauth, 1993 and Trauth et al., 1994).

During a previous work (Hudebine et al., 2002), two algorithms for molecular reconstruction have been developed in order to rebuild light cycle oil (LCO) gasoils from overall petroleum analyses. The first method was based on stochastic sampling of statistical distributions (SR), the other was a new approach, termed "reconstruction by entropy maximization" (REM). During validation, both methods have been applied to LCO gasoils because a larger number of analytical data was available for this type of feeds and the properties of the resulting mixture could thus be compared with the original data. The results have shown a reasonably good agreement, but they also illustrated some of the drawbacks of each method. Hence, coupling the SR and REM methods had been proposed. In the present work, this combination of both approaches will be tested on various LCO gasoils to illustrate the improvement of the coupled method compared to each approach taken separately.

GENERAL DESCRIPTION OF THE TWO-STEP RECONSTRUCTION ALGORITHM

In the present work, a two-step algorithm for molecular reconstruction was developed. In a first step, a large representative set of molecules is created, containing several thousands of species. This is done via a SR method. The second step in the algorithm aims at improving the composition of this set of molecules by adjusting the mole fractions of the various species via the REM method.

The "Stochastic Reconstruction" Step

The SR step creates an initial set of molecules, which is then successively modified until a mixture is obtained that mimics the properties of the experimental data. To achieve this, the SR method uses parametric distributions of structural blocks (polycyclic cores, rings, chains, etc.) in order to characterize the oil fraction to be reconstructed. The transition from a set of distributions into a molecule is performed N times by a Monte-Carlo sampling method, generating a mixture of N molecules. During the transition, a "building diagram" and "chemical rules" are applied to avoid the creation of unfeasible or unlikely molecules. The building diagram is a decision tree that describes the sampling steps of the different distributions in the correct order. The chemical rules are criteria that discard molecules based on thermodynamic or likelihood grounds. For each molecule, pure compound properties are also calculated, either directly by inspection of the structure (e.g. chemical formula, molecular weight, NMR), or numerically by group contribution methods (e.g. specific gravity, normal boiling point). The Monte-Carlo sampling creates a mixture of N molecules, each with a molar fraction of 1/N. This mixture is therefore called "equimolar", even if a same structure may have been generated several times. The average properties of this equimolar mixture are obtained through mixing rules and compared to the available analyses through an objective function. This objective function is then minimized by means of an optimization method, which modifies the parameters of the distributions. The basic algorithm of the SR step is depicted in Fig. 1.

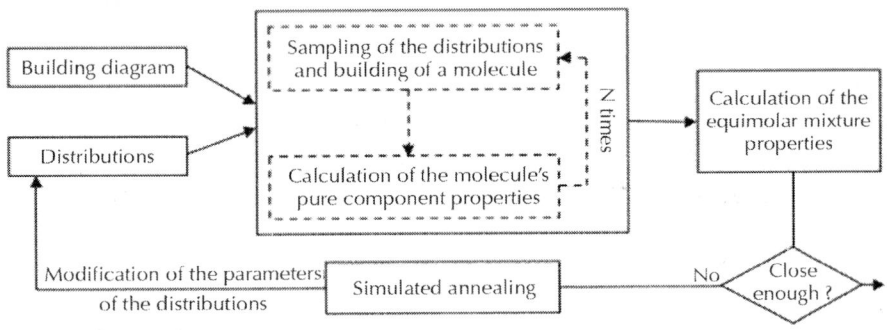

Figure 1: Flowchart of the stochastic reconstruction (SR) step.

The "Reconstruction by Entropy Maximization" Step

The second step of the algorithm starts from the set of N molecules defined previously and modifies their molar fractions in order to obtain a mixture whose properties are more closely aligned to those of the petroleum cut. To adjust the molar fractions, the model maximizes the information entropy, hence the name of the method. This criterion guarantees that a given molecule cannot be preferred to others if no information is available. In other words, without constraints or information, the distribution of the predefined set of molecules is uniform. When constraints are introduced, the uniform distribution is distorted in order to match this information. If the constraints are linear as a function of the molar fractions, the resolution of the problem is semi-algebraic and can be reduced to a maximization problem of a non-linear equation with *J* parameters, with *J* the number of constraints to verify (Hudebine, 2003). The basic algorithm of the REM step is shown in Fig. 2.

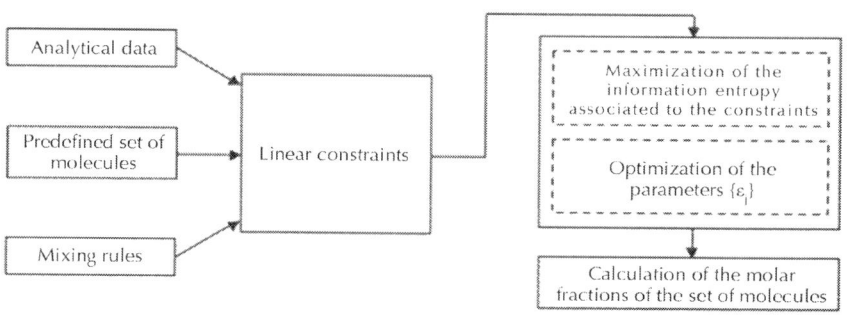

Figure 2: Flowchart of the reconstruction by entropy maximization (REM) step.

Advantages of the Two-step Algorithm

Each of the two steps described above could be used as a stand-alone reconstruction method. SR has previously been applied to rebuild various LCO gasoils (Hudebine et al., 2002 and Hudebine, 2003). The results have shown a good agreement between the properties of the reconstructed mixtures and the analytical data. However, some shortcomings also appeared, especially concerning the fit to some analyses. For example, in the mass spectrometry analysis, monocycloparaffins (C_nH_{2n}) are systematically underestimated, while the family of indanes/tetralins (C_nH_{2n-8}) is overestimated. This mismatch has been attributed to internal constraints of the SR method, which are implicitly contained in the molecule construction scheme. Consequently, the problem is intrinsic to the SR method

Chapter 4

Experimental Campaign, Modeling, and Sensitivity Analysis for the Molecular Distillation of Petroleum Residues 673.15K+

L. Zuñiga Liñan[a], N.M. Nascimento Lima[b,1], F. Manenti[c,2], M.R. Wolf Maciel[a,1], R. Maciel Filho[b,1], and L.C. Medina[d,3]

[a]University of Campinas, UNICAMP, School of Chemical Engineering, Separation Process Development Laboratory (LDPS), 13083-970, Campinas-SP, Brazil

[b]University of Campinas, UNICAMP, School of Chemical Engineering, Optimization, Project and Advanced Control Laboratory (LOPCA), 13083-970, Campinas-SP, Brazil

[c]Politecnico di Milano, CMIC Department "Giulio Natta", Piazza Leonardo da Vinci, Milano, Italy

[d]CENPES/PDP/TPAP/PETROBRAS, Centro de Pesquisas e Desenvolvimento da Petrobras, Rio de Janeiro, Brazil

ABSTRACT

This research activity proposes a sensitivity analysis of the molecular distillation process by focusing the attention on the response of the overall distillate flow rate under several conditions of distillation temperature and feed flow rate. Specific equations to characterize physicochemical properties of petroleum residues have been formulated by means of ASTM-based experimental campaigns combined with specific optimization techniques.

The steady state refining process simulator by Petrobras coupled with appropriate finite-difference methods is adopted for the simulation of a heated and extremely low-pressure falling film evaporator to separate a heavy residue 673.15 K+ of Gamma + Sigma crude oil. Numerical results are validated on the experimental points.

INTRODUCTION

Molecular distillation is a separation process based on the free transfer of molecules evaporated (unobstructed-path distillation) under high vacuum. The distance of transfer is comparable with the mean free path of the vapor molecules in the residual gas (short-path evaporation). The process operates at the lowest temperature and involves the least thermal decomposition as already discussed elsewhere (Hickman, 1943).

Between 1920 and 1940 the technique presented a revolutionary evolution, where those world's most plentiful raw materials considered "undistillable", such as the natural oils, fats, and waxes, were distillated by molecular distillation process.

In the middle 1920s C.R. Burch began experimentations in high-vacuum distillation (Hickman, 1943). He was one of the earliest workers to employ the Langmuir mercury condensation pump for producing the high vacuum in a still. Also, he examined the residue from petroleum refineries and demonstrated that a substantial proportion of this hitherto undistillable mixture could be distilled.

From vaseline he produced mobile liquid fractions of high molecular weights and exceedingly low vapor pressure. In the falling-film molecular distiller, the flowrate to be distilled is allowed to flow by gravity down a hot vertical surface on which it spreads in layer generally 1×10^{-4} to 1×10^{-3} m, thick according to viscosity and the flowrate itself. The principal unit of falling-film distiller is showed in Fig. 1.

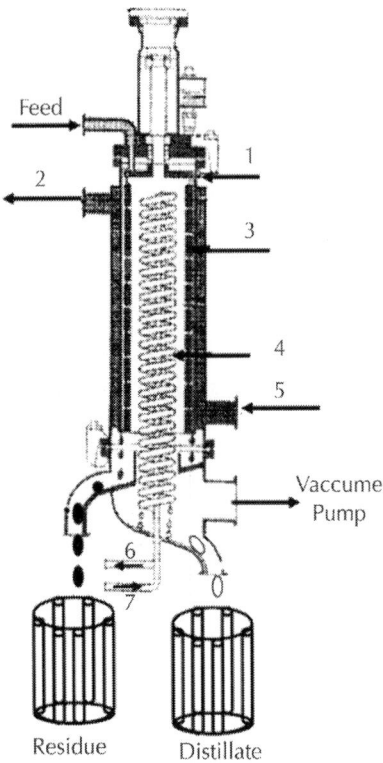

Figure 1: Scheme of falling-film molecular distillator used for the separation of atmospheric residue 673.15 K+ of Gamma + Sigma crude oil. 1 – rotating distribution plate; 2 – thermal fluid output; 3 – roller wiper; 4

– condenser; 5 – thermal fluid input; 6 – cooling fluid output; 7 – cooling fluid input; 8 – evaporator.

Since its construction, the equipment has been used especially to separate vitamins A and D in the Kodak laboratory, to recover carotenoids from palm oil (Batistella and Wolf Maciel, 1998), to recover vitamin E from soy oil (Batistella et al., 2002) and to purify thermally sensitive products (Chen et al., 2005). Recent works in LDPS/LOPCA/UNICAMP laboratories have demonstrated the performance of molecular distillation process to fractionate atmospheric and vacuum residues of crude oils at the laboratory scale (Maciel Filho et al., 2006 and Sbaite et al., 2006).

Nowadays, detailed modeling and sensitivity analysis may provide relevant pieces of information for the industrial scale-up of the molecular distillation process.

Several authors proposed mathematical models of molecular distillation process by reporting experimental distillation results for binary mixture and specific design configurations.

Kawala and Stephan (1989) developed the model and simulation of the molecular distillation process in a laminar falling-film unheated evaporator for the binary mixture of di-n-butyl-phthalate/di-n-butyl-sebacate.

Lutišan and Cvengroš (1995) defined a one-dimensional model of the molecular distillation based on the Monte Carlo simulation method to calculate particle velocities within the distillation space and to calculate some macroscopic features such as particle density, collision frequency, mean free path, and kinetic temperature.

Batistella and Wolf Maciel (1996) compared the performance of falling-film unheated and centrifugal evaporator for a binary mixture by using the model developed by Kawala and Stephan (1989).

Micov et al. (1997) presented a simplified model of molecular distillation that accounts for physical phenomena taking place in the liquid film and in the gas phase. Batistella and Wolf Maciel (1998) described a technical process to recover carotenoids from palm oil through transesterification and molecular distillation by means of falling-film and centrifugal molecular distillers. Also, they compared the experimental and simulation results.

Cvengroš et al. (2000) investigated the behavior of film surface temperature along the height of the evaporator under steady-state conditions at different values of the peripheral liquid load of evaporating cylinder, of the temperature differences between evaporation surface temperature and entrance liquid temperature, and of the temperature of evaporation surface.

Batistella et al. (2000) incorporated a rigorous model for the vapor phase in the DISMOL software (Batistella, 1999) to evaluate industrial applications of molecular distillation process. DISMOL is able to foresee the behavior of the molecular distillation in terms of several factors such as design of the molecular distillators, pressure of the system, and condensation temperature to quote a few.

Cvengroš et al. (2001) developed an experimental study for wiped film evaporator to investigate the function of roller wiper on the liquid flow regimes at different liquid loads of the evaporato s perimeter and different wiper peripheral speeds. Batistella et al. (2002) showed the potentiality of molecular distillation process for recovering vitamin E from vegetal oils. Also, the author compared the performance of the falling film and centrifugal distillers.

Lutišan et al. (2002) developed a mathematical model of evaporation from the liquid film surface at high vacuum to evaluate differences between the laminar and the turbulent regimes in the film depending on the evaporato s load and temperature.

With the aim to provide information on the choice of adequate operating conditions, Xubin et al. (2005)presented a mathematical model that describes heat and mass transfer in the film of the evaporating liquid on the heated cylinder evaporator and in the gas phase in the distillation gap by considering a binary system in the presence of an inert gas.

Sales-Cruz and Gani (2006) presented a generalized short-path evaporation model, (based on Kawala and Stepha s model), with a corresponding systematic simulation and solution strategy for various types of design/analysis problems. The author presented two case studies: the purification of a reaction mixture containing glycerol, mono-, di- and triglycerides and the recovery of a pharmaceutical product from a six-component mixture.

One of the problems arising when trying to predict the performance of those processes that involve petroleum products is the lack of experimental data and of opportunely detailed and field-proven models. This problem is emphasized for petroleum residues that cannot be removed by the atmospheric distillation (Merdrignac and Espinat, 2007).

For the sake of clarity, one of the most common petroleum residues is the so-called AR (Atmospheric Residue) pointing out the material at the bottom of the atmospheric distillation column (ASTM D 2892 method) which has an Atmospheric Equivalent Boiling Point (AEBP) higher than 673.15 K (in the following, 673.15 K+).

Petroleum residues are highly complex products which have high viscosity, significant content of heteroatoms and high molecular weights and these characteristics are directly related to the major presence of compounds with complex structures, typical of resins and asphaltenes (Merdrignac and Espinat, 2007).

There are many research papers in the literature proposing mathematical models and simulation results of the molecular distillation process for binary mixtures, in condition of low or moderate temperatures in adiabatic evaporator. Conversely, looking forward an industrial scale up of molecular distillation, other important conditions such as high process temperatures, very high vacuum pressure, and low flow rate must be analyzed. Also, the limitation of binary mixtures must be removed too.

Thus, the present research activity mainly deals with these objectives by considering the atmospheric residue Gamma + Sigma 673.15 K+ as a multicomponent mixture consisting of 6 pseudocomponents, which were characterized by availing from Petrobras simulator (PETROX) and from correlations for heavy fractions (Niederberger et al., 2005, Pedersen et al., 1984 and Poling et al., 2004). To perform the molecular distillation of the selected petroleum residue, a falling-film molecular distiller with roller wiper system and constant heated in the evaporator wall is adopted in the present work.

EXPERIMENTAL CAMPAIGN

Atmospheric Residue Gamma + Sigma 673.15 K+

A simulation of the complete fractionation of crude oil Gamma + Sigma was performed by PETROX. This simulation enabled to predict the total number of evaporated components during the conventional distillation of petroleum. The results showed a total number of 102 components which comprise from gases and vapor (water, nitrogen, methane) to components that evaporate at temperatures higher 1000 K. From the complete list, the evaporated components from 645.15 K to 1084.15 K were considered as being the constitutive of atmospheric residue. The High-Temperature Simulated Distillation (HTDS) by Gas Chromatography (GC) of atmospheric residue enabled to define the Initial Boiling Point (IBP) of most volatile component of the residue (Zuñiga et al., 2010a and Zuñiga et al., 2010b). HTDS measurements were made by the Research Center of Petrobras-Brazil (CENPES/Petrobras). The physicochemical characterization of the total components present in the residue, such as AEBP, molecular weight, specific gravity, API gravity and the percentage vaporized (%, v/v) are listed in Table 1. The list shows a total of 61 components which are grouped into 6 pseudocomponents, according to similarities of their properties and specifically the API gravity. Each group pseudocomponent is denoted by a letter: "a", "b", "c", "d", "e" and "f". The pseudocomponents present different features and different amount of components involved: the pseudocomponent "a" is the most volatile one and it consists of 5 components, whereas the pseudocomponent "f" is the heaviest one as well as the largest since it includes 19 components.

Table 1: Estimation of properties for the pseudocomponents adopted to characterize the atmospheric residue Gamma + Sigma 673.15 K+ and initial molar compositions of the inlet flow rate. $m_0 = 2.44 \times 10^{-3}$ kg s^{-1} is considered as basic value for the calculations

Pseudocomponent	Component	AEBP (K)	M_w	SG	API	% (v/v)	T_{AVBP^q} (K)	SG_q	M_{wq}	x_{q0}
"a"	1	645.15	298	0.9273	21.0935	1.0418	657.7	0.9335	311	0.1312
	2	651.15	304	0.9304	20.5833	1.0620				
	3	657.15	311	0.9336	20.0668	1.0817				
	4	663.15	318	0.9366	19.5804	1.1006				
	5	670.15	325	0.9395	19.1097	1.1594				
"b"	6	675.15	332	0.9420	18.7164	1.0595	690.2	0.9474	351	0.1572
	7	680.15	338	0.9440	18.3911	0.9566				
	8	685.15	344	0.9460	18.0780	1.0093				
	9	690.15	350	0.9477	17.8160	1.0205				
	10	695.15	357	0.9492	17.5656	1.0674				
	11	700.15	363	0.9507	17.3363	1.0631				
	12	705.15	370	0.9520	17.1276	1.0586				

Experimental Campaign, Modeling, and Sensitivity Analysis....　91

"c"	13	710.15	377	0.9533	16.9308	1.1041	733.7	0.9565	414	0.2177
	14	715.15	384	0.9540	16.8218	1.1127				
	15	720.15	392	0.9547	16.7152	1.1453				
	16	725.15	399	0.9553	16.6140	1.0738				
	17	730.15	406	0.9560	16.5187	1.0008				
	18	734.15	413	0.9566	16.4196	1.0302				
	19	739.15	421	0.9573	16.3190	1.0339				
	20	743.15	428	0.9579	16.2170	1.0371				
	21	748.15	435	0.9586	16.1138	1.0396				
	22	753.15	443	0.9588	16.0857	1.0412				
	23	757.15	451	0.9591	16.0416	1.0417				
"d"	24	762.15	458	0.9595	15.9794	1.0408	777.2	0.9617	484	0.1211
	25	766.15	466	0.9600	15.8993	1.0388				
	26	771.15	474	0.9606	15.8033	1.0192				
	27	776.15	483	0.9614	15.6819	1.0868				
	28	781.15	493	0.9624	15.5319	1.1531				
	29	787.15	503	0.9635	15.3636	1.1305				
	30	792.15	513	0.9647	15.1757	1.1239				

92 Handbook of Petroleum Analysis

"e"								
31	798.15	523	0.9661	14.9656	1.1164	830.7	0.9764	586	0.1706
32	803.15	533	0.9677	14.7175	1.1081				
33	809.15	544	0.9695	14.4590	1.099				
34	814.15	554	0.9712	14.1908	1.0892				
35	820.15	565	0.9730	13.9234	1.0402				
36	826.15	577	0.9749	13.6389	1.0633				
37	832.15	589	0.9770	13.3370	1.0858				
38	838.15	602	0.9790	13.0386	1.0349				
39	844.15	615	0.9810	12.7339	1.0218				
40	851.15	628	0.9834	12.3909	1.0079				
41	857.15	642	0.9858	12.0376	0.9935				
42	863.15	655	0.9882	11.6908	0.9786				

and cannot be eliminated easily. As a Monte-Carlo technique is used, the SR method is also computationally intensive.

As opposed to the SR method, the computational effort of the REM method is much smaller. Moreover, this method allows to obtain a mixture whose properties are very close to the analytical data. However, this approach also exhibits some difficulties when used as a stand-alone method. Firstly, the method strongly depends on the initial set of molecules. That is why the molecules must be chosen with care, i.e. corresponding to the type of feedstock studied. Secondly, the search of a solution gets increasingly more difficult when the properties of the initial set of molecules (which is considered to be equimolar) are too far from the analytical data. In extreme cases, if the initial mixture is very different, the dispersion of the molar fractions is so important that the optimized mixture is composed of a very small number of highly concentrated compounds, while the remaining compounds have a near-zero mole fraction. The REM method is also very sensitive to inconsistencies in the analytical data, in contrast to the SR method that has intrinsic data reconciliation properties.

To summarize the individual approaches and underline their advantages and drawbacks, the SR method is a very constrained method that yields an equimolar mixture, but only roughly approaches the analytical data. On the other hand, the reconstructed molecules are always typical of the studied oil fraction due to the use of specific building rules based on analytical and expert knowledge. Opposed to the SR method, the REM method is a finishing method that allows to find a mixture with properties that are very close to analytical data, but it needs a particularly well-defined initial set of molecules. For these reasons, coupling both methods seems an excellent solution to eliminate their respective drawbacks.

APPLICATION OF THE TWO-STEP RECONSTRUCTION ALGORITHM TO A LCO GASOIL

Construction of the Initial Set of Molecules by the SR Step

The SR method can be applied to any type of feedstock with a limited number of modifications. In the case of LCO gasoil reconstruction, only the type of the distributions and the building diagram must be specified.

The choice of the set of distributions is very important and needs some expert knowledge of the chemical structure of LCO gasoils. For example, due to the cut point of classical LCO gasoils, a representative molecule cannot have more than one polycyclic core. A histogram with two bins, 0 or 1, is then enough to describe this structural part of gasoil molecules. Analogously, the maximum length of the alkyl chains is necessarily limited because of the cracking reactions in the catalytic cracking process. The distribution of the alkyl chain length can be described by a histogram with only 3 possibilities: 1, 2 or 3 carbon atoms. Table 1 lists the 9 distributions and the 15 associated parameters used to build LCO gasoils.

Table 1: Distributions used for the reconstruction of LCO gasoils by the SR step

		Initial value	Optimized value
Distribution 1	Probability to have 0 core:	0.200	0.283
Number of poly-cyclic cores	Probability to have 1 core:	Complement to 1	Complement to 1

	Distribution 2 Number of aromatic rings per core	Probability to have 0 ring:	0.400	0.178
		Probability to have 1 ring:	0.200	0.223
		Probability to have 2 ring:	0.200	0.451
		Probability to have 3 ring:	Complement to 1	Complement to 1
	Distribution 3 Number of naphtenic rings per core	Probability to have 0 ring:	0.400	0.476
		Probability to have 1 ring:	0.200	0.327
		Probability to have 2 ring:	0.200	0.076
		Probability to have 3 ring:	Complement to 1	Complement to 1
	Distribution 4 Number of thiophenic rings per core	Probability to have 0 ring:	0.800	0.772
		Probability to have 1 ring:	Complement to 1	Complement to 1
	Distribution 5 Does an aromatic CH accept a chain?	Probability for "no":	0.700	0.780
		Probability for "yes":	Complement to 1	Complement to 1
	Distribution 6 Does a naphtenic CH_2 accept a chain?	Probability for "no":	0.700	0.428
		Probability for "yes":	Complement to 1	Complement to 1
	Distribution 7 Type of the chain	Probability to have a methyl:	0.700	0.314
		Probability to have an ethyl:	0.150	0.481
		Probability to have an isopropyl:	Complement to 1	Complement to 1

Distribution 8	Value for α:	2.000	2.088
Size of the paraffins	Value for β:	4.000	4.438
Distribution 9	Probability for "no":	0.700	0.996
Do a paraffinic CH2 accept a chain?	Probability for "yes":	Complement to 1	Complement to 1

To define the hierarchic relationships between distributions and identify the sampling steps, a specific building diagram is also needed. This operation is relatively simple for LCO gasoils. For each molecule to rebuild, the distribution 1 is sampled first. If the number of cores is equal to 0, the molecule is a paraffin and distribution 8 is used to determine its length. If the number of cores is equal to 1, distributions 2–4 are successively used to select the number of aromatic, cyclohexanic and thiophenic rings in the core. Depending on the type and number of rings, a 2D structure is chosen from a library of predefined cores. After that, each aromatic and naphtenic carbon atom is tested to define whether an alkyl chain is inserted or not (distributions 5 and 6). Finally, if a chain is accepted, distribution 7 is used to determine its length.

The above-described operations are applied N times so as to obtain a mixture of N molecules that is statistically representative of the distributions. For each molecule, the elemental composition, molecular weight, mass spectrum, ^1H NMR spectrum and ^{13}C NMR spectrum are easily obtained. For the specific gravity and the normal boiling point, novel group contribution methods are used (Hudebine, 2003). The properties of the equimolar mixture are subsequently calculated by linear mixing rules (elemental analysis, average molecular weight, average specific gravity, simulated distillation, mass spectrometry, ^1H and ^{13}C NMR analyses). In the case of the average specific gravity and the distillation, this is equivalent to the assumption of an ideal mixture behavior. Finally, the 15 distribution parameters are modified in order to adjust the equimolar mixture until the desired mixture properties are found. The difference between experimental and calculated data is mathematically described by an objective function, which is minimized by a simulated annealing algorithm.

The SR step was applied to a LCO gasoil whose properties are given in Table 2. For this sample, the available analyses were the elemental analysis, simulated distillation, mass spectrometry, ^{13}C NMR analysis and ^{1}H NMR analysis. The number of molecules to be sampled was set to 5000. The equimolar mixture obtained after simulated annealing is called "Set L". Its calculated mixture properties are also given in Table 2. Comparison to the experimental analyses shows a good agreement between the properties of the generated mixture and the analytical data of the LCO. Table 1 provides the initial and optimized parameters of the distributions used.

Table 2: Properties of rebuilt mixtures compared to the analytical data for a LCO gasoil

Petroleum analyses			Analytical data	Set L	Set L+
Global analyses	Carbon content	(wt%)	88.05	87.80	88.06
	Hydrogen content	(wt%)	10.42	10.69	10.42
	Sulfur content	(wt%)	1.53	1.51	1.52
	Specific gravity	(g/cm³)	0.9407	0.9333	0.9537
Simulated distillation	Initial boiling point	(°C)	135	134	134
	5 wt%	(°C)	207	171	153
	10 wt%	(°C)	231	208	223
	20 wt%	(°C)	264	245	250
	30 wt%	(°C)	286	269	281
	40 wt%s	(°C)	305	298	302
	50 wt%	(°C)	323	327	318
	60 wt%	(°C)	342	356	336
	70 wt%	(°C)	361	377	357
	80 wt%	(°C)	381	396	376

	90 wt%	(°C)	403	416	401
	95 wt%	(°C)	418	433	443
	Final boilingpoint	(°C)	454	459	458
Mass spectrometry	C_nH_{2n+2}	(wt%)	13.1	18.6	13.1
	C_nH_{2n}	(wt%)	9.9	6.0	9.9
	$C_nH_{2n(-2,-4)}$	(wt%)	5.4	6.3	5.4
	C_nH_{2n-6}	(wt%)	7.1	4.3	7.1
	C_nH_{2n-8}	(wt%)	7.0	7.1	7.0
	C_nH_{2n-10}	(wt%)	2.0	2.0	2.0
	C_nH_{2n-12}	(wt%)	16.7	18.1	16.7
	C_nH_{2n-14}	(wt%)	10.7	9.6	10.7
	C_nH_{2n-16}	(wt%)	6.9	10.0	6.9
	C_nH_{2n-18}	(wt%)	7.0	7.6	7.0
	C_nH_{2n-20} and more	(wt%)	5.1	0.3	5.1
	$C_nH_{2n-10}S$	(wt%)	4.5	2.4	4.5
	$C_nH_{2n-16}S$	(wt%)	4.6	7.7	4.6
^{13}C NMR	Saturated CH_3	(mol%)	16.7	19.0	17.5
	Saturated CH_2	(mol%)	32.0	26.7	29.3
	Saturated CH	(mol%)	7.7	10.0	7.5
	Saturated C	(mol%)	0.4	0.0	0.0
	Aromatic CH	(mol%)	22.3	24.8	22.4
	Condensed aromatic C	(mol%)	7.6	7.7	8.0
	Substituted aromatic C	(mol%)	13.3	11.8	15.3
^{1}H NMR	Diaromatic H	(mol%)	13.6	14.7	13.8
	Monoaromatic H	(mol%)	2.3	2.4	2.1
	H (type alpha)	(mol%)	23.1	13.9	22.9
	H (type beta)	(mol%)	43.9	48.3	45.9
	H (type gamma)	(mol%)	17.1	20.7	15.3

Improvement of the Initial Set of Molecules by the REM Step

To further improve the representativeness of this set of molecules, the REM method is used to determine the molar fractions of each of the molecules in the set. Table 2 also lists the calculated properties of this new mixture ("Set L+"), which is now no longer equimolar as each molecule has its own molar fraction that has been defined via the information entropy criterion. As can be seen, the REM step only adds small improvements on more global analyses, such as the elemental analysis. However, on more specific and detailed analyses, such as NMR and mass spectrometry, the REM step allows to refine the agreement between the mixture properties and the analytical data in a significant manner.

RECONSTRUCTION OF VARIOUS LCO GASOILS USING A MODIFIED TWO-STEP RECONSTRUCTION ALGORITHM

As illustrated above, the two-step reconstruction algorithm can be applied to LCO gasoils to generate, for each sample, a detailed mixture that adequately mimics the properties of the gasoil to be represented. However, as the first step of this reconstruction method uses a Monte-Carlo approach to generate its initial set of molecules, it still remains a computationally intensive algorithm. A modified two-step reconstruction algorithm was therefore derived and may be applied when several LCO gasoils need to be represented.

Description of the Approach

A reference LCO is first reconstructed through the above-described two-step procedure. Hence, the resulting molecule set can be considered as a reference mixture that is typical for LCO gasoils. To reconstruct other LCO gasoils, only the REM step is employed. It starts from the reference mixture (instead of a specific Monte-Carlo generated mixture) to obtain a new mixture whose properties are very close to those of the LCO gasoil to represent. This mixture will now contain the same molecules as the reference mixture but with modified molar fractions. Fig. 3 summarizes this modified two-step reconstruction algorithm.

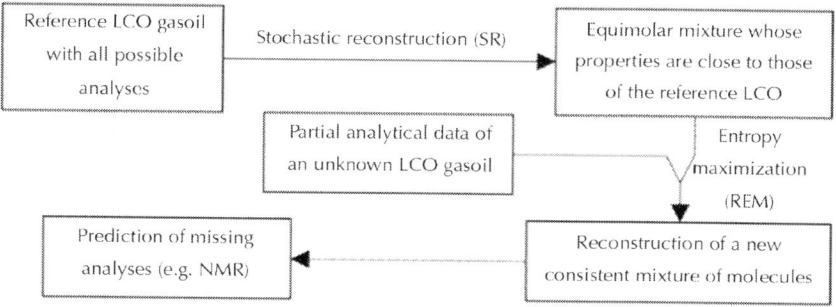

Figure 3: Flowchart of the modified two-step reconstruction algorithm.

To be representative of many LCO gasoils, a gasoil with the widest boiling point range was chosen as reference gasoil in order to be able to encompass all gasoils. The reference gasoil chosen corresponds to the gasoil that was already presented in Table 2. The Set L+ from Table 2 is not equimolar, however, a condition necessary for the information entropy criterion. To return to an equimolar set of molecules, it is necessary to multiply the molar fraction x_i of each molecule i by a constant factor K (50 000 for example) and to insert this molecule $K \cdot x_i$ times, rounded to the nearest integer, into the new mixture set. Hence, an "equimolar" set of 50 000 molecules is obtained whose properties are exactly equal to those of the non-equimolar mixture of 5000 molecules.

This new set constitutes an excellent starting point. Indeed, the constructed molecules are representative of LCO gasoil molecules thanks to the building rules used by the SR step. The properties of the corresponding equimolar mixture are very close to the analytical data of the reference LCO gasoil and are, by definition, also typical of analytical data of any LCO gasoil. Consequently, this database of 50 000 molecules will be used to rebuild other LCO gasoils from their analytical data by the REM step.

To further illustrate the information entropy maximization step, the equations will now be developed. First, the entropic criterion to be maximized is given by

$$H = -\sum_{i=1}^{50\,000} x_i \cdot \ln x_i, \quad (1)$$

where x_i is the molar fraction of the ith molecule of the set of 50 000 molecules.

Associated with this criterion, various constraints must be introduced. The first one allows to ensure the normalization of the molar fractions, which can be written as

$$\sum_{20\,000}^{i=1} x_i = 1^\circ \quad (2)$$

The other constraints arise from the analytical data to be fitted. As the mixing rules are considered to be linear, these constraints can be written as follows:

$$\sum_{i=1}^{50\,000} x_i \cdot f_{i,j} = f_j^{exp} \quad \forall j \in J, \quad (3)$$

Where J is the set of constraints, $f_{i,j}$ is the value of the property of the ith molecule associated to the j th constraint and f_j^{exp} is the experimental value of the analysis associated to the j th constraint. The value of $f_{i,j}$ is determined for each analysis. For example, in the case of the mass spectrometry analysis, $f_{i,j}=0$ when molecule i does not belong to mass spectrometry class m corresponding to constraint j and $f_{i,j}=M_i/M_{mixture}$ when molecule i belongs to mass spectrometry class m corresponding to constraint j. For the simulated distillation analysis, $f_{i,j}=0$ when the boiling point of molecule i is above the experimental boiling $BP_{exp}(m)$ and $f_{i,j}=M_i/M_{mixture}$ when the boiling point of molecule *i* is lower than the experimental boiling $B_{Pexp}(m)$. In these formulas, M_i is the molecular weight of molecule *i* and $M_{mixture}$ is the average molecular weight of the mixture obtained from $\Sigma(x_i . M_i)$.

When these constraints are introduced into the entropy criterion, Eq. (1) is modified and the new equation to maximize becomes

$$\zeta = -\sum_{i=1}^{50\,000} x_i \cdot \ln x_i + \mu \cdot \left(1 - \sum_{i=1}^{50\,000} x_i \right) + \sum_{j=1}^{J} -\frac{1}{2} \cdot \left(\frac{f_j^{exp} - \sum_{i=1}^{50\,000} x_i \cdot f_{i,j}}{\sigma_j} \right)^2, \quad (4)$$

Where μ is a Lagrange multiplier and σ_j is a standard deviation associated to the *j*th constraint.

Finally, the algebraic solution of the maximization problem is the following:

$$x_i = \frac{z_i}{Z}, \quad Z = \sum_{i=1}^{50000} z_i, \quad z_i = \exp\left(\sum_{j=1}^{J} \frac{f_{i,j}}{\sigma_j} \cdot \varepsilon_j \right), \quad (5)$$

where the vector of parameters $\{\varepsilon_j\}$ is calculated to maximize the non-linear Eq. (6):

$$E = \ln Z + \sum_{j=1}^{J} \left(\frac{1}{2} \cdot \varepsilon_j^2 - \frac{f_j^{exp}}{\sigma_j} \cdot \varepsilon_j \right) \quad (6)$$

This last equation is a multidimensional non-linear function. Its maximization can be performed by a simple conjugate gradient method. Once the parameters $\{\varepsilon_j\}$ are optimized, the various molar fractions x_i can be easily calculated from Eq. (5). The mixture is now reconstructed.

Validation of the Approach

Nine LCO gasoils have been reconstructed using the modified two-step reconstruction algorithm. The analyses used for the molecular reconstruction were the simulated distillation and mass spectrometry analyses. These analyses have been chosen because they give some information in term of molecule size (distillation) and chemical structure (mass spectrometry). To validate the reconstruction, the best solution would be to compare the rebuilt mixtures to the real gasoils molecule by molecule. However, this is impossible as there are no analyses that allow to identify and quantify the individual molecules present in gasoils. As a substitute for a molecule-based validation, properties that were not used during the reconstruction (elemental analysis, specific gravity, molecular weight, ^{1}H NMR, ^{13}C NMR) were predicted from the rebuilt mixtures and compared to their corresponding experimental values. Fig. 4 gives the parity plots for the predicted specific gravity, molecular weight, carbon content, hydrogen content, sulfur content, as well as for the distribution of hydrogen atoms by NMR type and for the distribution of carbon atoms by NMR type.

The results show a good agreement between the predicted and experimental analyses. Two problems have nevertheless been encountered. The first one concerns the specific gravity. If the ideal mixture hypothesis is directly used, a constant bias is observed between analytical data and calculated mixtures properties. This bias has been evaluated as +0.0113 g/cm^3 and can be subtracted during the mixture calculations. The second problem concerns the average molecular weight. The estimated molecular weights are always smaller than the "experimental" molecular weights, which are obtained from a correlation based on specific gravity and simulated distillation (API, 1987). The average bias equals –9 g / mol and is probably due to the hypothesis that the simulated distillation perfectly separates molecules by increasing normal boiling points. In practice, however, chromatographic columns are not strictly apolar and separate to some extent according to the polarity of the molecule. For example, the simulated distillation boiling point for naphthalene is 11 C lower than its normal boiling point, while for anthracene this difference already amounts to 36 C (ASTM D2887, 2002). Hence, the reconstructed polynuclear molecules are slightly smaller than their actual counterparts, resulting in a slightly lower average molecular weight. For the other analyses (elemental analysis, NMR analyses), the results are very consistent and, for most of the evaluated properties, lie within the analytical confidence intervals. Hence, the coupled SR/REM molecular reconstruction scheme can be used as alternative for the analyses themselves. This is particularly the case for the NMR analyses, which are expensive and consequently hardly used.

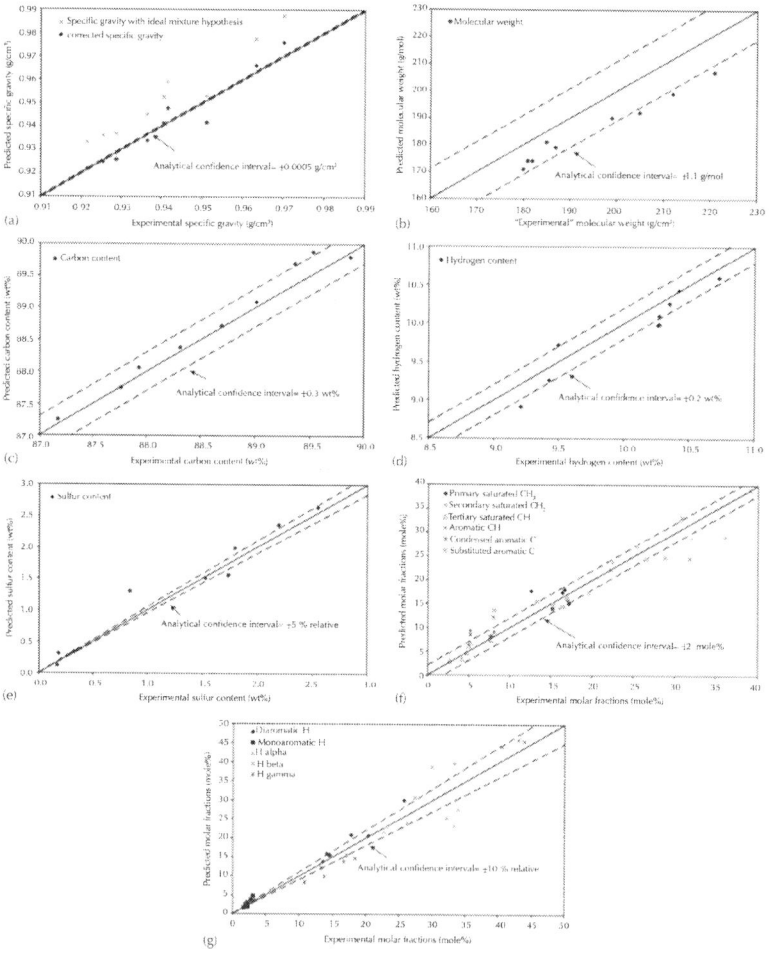

Figure 4: Parity plots between analytical data and estimated properties of the mixtures reconstructed by the coupled SR/REM algorithm. (a) specific gravity, (b) molecular weight, (c) carbon content, (d) hydrogen content, (e) sulfur content, (f) ^{13}C NMR and, (g) ^{1}H NMR.

CONCLUSIONS

Stochastic reconstruction (SR) is a method that yields molecular sets whose mixture properties are very close to those of the petroleum fraction that is represented, but the achievement of a perfect-fit solution is perturbed by the building algorithm, as it strongly constrains the structure of the generated molecules.

Reconstruction by entropy maximization (REM) is CPU efficient and provides a very good agreement between simulated and experimental properties, but the mixture can be very difficult to reconstruct and inconsistent if the initial molecular set is not correct or very different from the feedstock type.

For these reasons, a two-step reconstruction algorithm coupling both methods has been proposed and two variants have been developed and validated. The idea consists in using a SR step to build a reference mixture. This set of molecules thus obtained can be used in a second step either to improve its representativeness or to rebuild similar petroleum fractions via the REM method. To validate the latter approach, nine LCO gasoils have been reconstructed with this technique from their simulated distillation and mass spectrometry analysis. The results in terms of analysis prediction have shown a very good agreement. Consequently, this two-step molecular reconstruction algorithm can be used to predict missing analytical data or to generate molecular mixtures that can serve as input to detailed kinetic models.

REFERENCES

1. Allen, D.T., Liguras, D.K., 1991. Structural models of catalytic cracking chemistry: a case study of a group contribution approach to lumped kinetic modeling. In: Sapre, A.V., Krambeck, F.J. (Eds.), Chemical Reactions in Complex Mixtures, Mobil Workshop. Van Nostrand Reinhold, New York.

2. API, 1987. API procedure 2B2.1 for estimating the molecular weight of a petroleum fraction. API Technical Handbook.
3. ASTMD2887, 2002. Standard test method for boiling range distribution of petroleum fractions by gas chromatography. Annual Book of ASTM Standards.ASTM, Philadelphia.
4. Faulon, J.L., Vandenbroucke, M., Drappier, J.M., Behar, F., Romero, M., 1990. Modélisation des Structures Chimiques des Macromolécules Sédimentaires: le logiciel XMOL. Revue de l'Institut Français du Pétrole 45 (2), 161–180.
5. Hirsch, E., Altgelt, K.H., 1970. Integrated structural analysis. A method for the determination of average structural parameters of petroleum heavy ends. Analytical Chemistry 42 (12), 1330–1339.
6. Hudebine, D., 2003. Reconstruction moléculaire de coupes pétrolières. Ph.D. Thesis, Ecole Nationale Supérieure de Lyon.
7. Hudebine D., Vera, C., Wahl, F., Verstraete, J., 2002. Molecular representation of hydrocarbon mixtures from overall petroleum analyses. A.I.Ch.E Spring Meeting, New Orleans, March 10–14.
8. Neurock, M., 1992. A computational chemical reaction engineering analysis of complex heavy hydrocarbon reaction systems. Ph.D. of the University of Delaware.
9. Neurock, M., Nigam, A., Trauth, D.M., Klein, M.T., 1994. Molecular representation of complex hydrocarbon feedstocks through efficient characterization and stochastic algorithms. Chemical Engineering Science 49 (24), 4153–4177.
10. Quann, R.J., Jaffe, S.B., 1996. Building useful models of complex reaction systems in petroleum refining. Chemical Engineering Science 51 (10), 1615–1635.
11. Speight, J.G., 1970. A structural investigation of the constituents ofathabasca bitumen by proton magnetic resonance spectroscopy. Fuel 49 (1), 76–90.
12. Trauth, D.M., 1993. Structure of complex mixtures through characterization, reaction, and modeling. Ph.D. of the University of Delaware.

13. Trauth, D.M., Stark, S.M., Petti, T.F., Neurock, M., Klein, M.T., 1994. Representation of the molecular structure of petroleum resid through characterization and Monte Carlo modeling. Energy and Fuels 8 (3), 576–580.
14. Zhang, Y., 1999. A molecular approach for characterization and property predictions of petroleum mixtures with applications to refinery modelling. Ph.D. of the University of Manchester.

Chapter 4

Experimental Campaign, Modeling, and Sensitivity Analysis for the Molecular Distillation of Petroleum Residues 673.15K+

L. Zuñiga Liñan[a], N.M. Nascimento Lima[b,1], F. Manenti[c,2], M.R. Wolf Maciel[a,1], R. Maciel Filho[b,1], and L.C. Medina[d,3]

[a]University of Campinas, UNICAMP, School of Chemical Engineering, Separation Process Development Laboratory (LDPS), 13083-970, Campinas-SP, Brazil

[b]University of Campinas, UNICAMP, School of Chemical Engineering, Optimization, Project and Advanced Control Laboratory (LOPCA), 13083-970, Campinas-SP, Brazil

[c]Politecnico di Milano, CMIC Department "Giulio Natta", Piazza Leonardo da Vinci, Milano, Italy

[d]CENPES/PDP/TPAP/PETROBRAS, Centro de Pesquisas e Desenvolvimento da Petrobras, Rio de Janeiro, Brazil

ABSTRACT

This research activity proposes a sensitivity analysis of the molecular distillation process by focusing the attention on the response of the overall distillate flow rate under several conditions of distillation temperature and feed flow rate. Specific equations to characterize physicochemical properties of petroleum residues have been formulated by means of ASTM-based experimental campaigns combined with specific optimization techniques.

The steady state refining process simulator by Petrobras coupled with appropriate finite-difference methods is adopted for the simulation of a heated and extremely low-pressure falling film evaporator to separate a heavy residue 673.15 K+ of Gamma + Sigma crude oil. Numerical results are validated on the experimental points.

INTRODUCTION

Molecular distillation is a separation process based on the free transfer of molecules evaporated (unobstructed-path distillation) under high vacuum. The distance of transfer is comparable with the mean free path of the vapor molecules in the residual gas (short-path evaporation). The process operates at the lowest temperature and involves the least thermal decomposition as already discussed elsewhere (Hickman, 1943).

Between 1920 and 1940 the technique presented a revolutionary evolution, where those world's most plentiful raw materials considered "undistillable", such as the natural oils, fats, and waxes, were distillated by molecular distillation process.

In the middle 1920s C.R. Burch began experimentations in high-vacuum distillation (Hickman, 1943). He was one of the earliest workers to employ the Langmuir mercury condensation pump for producing the high vacuum in a still. Also, he examined the residue from petroleum refineries and demonstrated that a substantial proportion of this hitherto undistillable mixture could be distilled.

From vaseline he produced mobile liquid fractions of high molecular weights and exceedingly low vapor pressure. In the falling-film molecular distiller, the flowrate to be distilled is allowed to flow by gravity down a hot vertical surface on which it spreads in layer generally 1×10^{-4} to 1×10^{-3} m, thick according to viscosity and the flowrate itself. The principal unit of falling-film distiller is showed in Fig. 1.

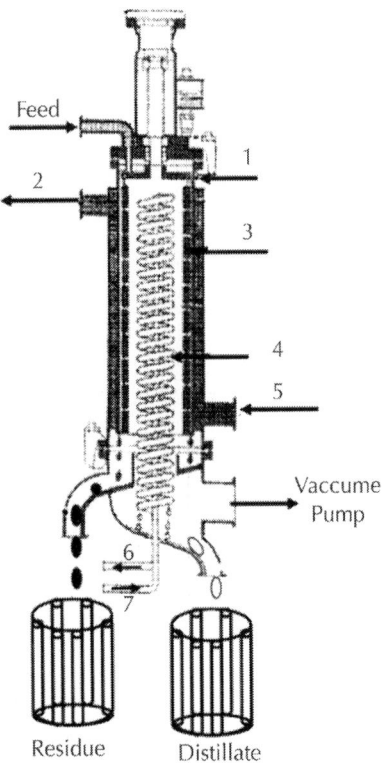

Figure 1: Scheme of falling-film molecular distillator used for the separation of atmospheric residue 673.15 K+ of Gamma + Sigma crude oil. 1 – rotating distribution plate; 2 – thermal fluid output; 3 – roller wiper; 4

– condenser; 5 – thermal fluid input; 6 – cooling fluid output; 7 – cooling fluid input; 8 – evaporator.

Since its construction, the equipment has been used especially to separate vitamins A and D in the Kodak laboratory, to recover carotenoids from palm oil (Batistella and Wolf Maciel, 1998), to recover vitamin E from soy oil (Batistella et al., 2002) and to purify thermally sensitive products (Chen et al., 2005). Recent works in LDPS/LOPCA/UNICAMP laboratories have demonstrated the performance of molecular distillation process to fractionate atmospheric and vacuum residues of crude oils at the laboratory scale (Maciel Filho et al., 2006 and Sbaite et al., 2006).

Nowadays, detailed modeling and sensitivity analysis may provide relevant pieces of information for the industrial scale-up of the molecular distillation process.

Several authors proposed mathematical models of molecular distillation process by reporting experimental distillation results for binary mixture and specific design configurations.

Kawala and Stephan (1989) developed the model and simulation of the molecular distillation process in a laminar falling-film unheated evaporator for the binary mixture of di-n-butyl-phthalate/di-n-butyl-sebacate.

Lutišan and Cvengroš (1995) defined a one-dimensional model of the molecular distillation based on the Monte Carlo simulation method to calculate particle velocities within the distillation space and to calculate some macroscopic features such as particle density, collision frequency, mean free path, and kinetic temperature.

Batistella and Wolf Maciel (1996) compared the performance of falling-film unheated and centrifugal evaporator for a binary mixture by using the model developed by Kawala and Stephan (1989).

Micov et al. (1997) presented a simplified model of molecular distillation that accounts for physical phenomena taking place in the liquid film and in the gas phase. Batistella and Wolf Maciel (1998) described a technical process to recover carotenoids from palm oil through transesterification and molecular distillation by means of falling-film and centrifugal molecular distillers. Also, they compared the experimental and simulation results.

Cvengroš et al. (2000) investigated the behavior of film surface temperature along the height of the evaporator under steady-state conditions at different values of the peripheral liquid load of evaporating cylinder, of the temperature differences between evaporation surface temperature and entrance liquid temperature, and of the temperature of evaporation surface.

Batistella et al. (2000) incorporated a rigorous model for the vapor phase in the DISMOL software (Batistella, 1999) to evaluate industrial applications of molecular distillation process. DISMOL is able to foresee the behavior of the molecular distillation in terms of several factors such as design of the molecular distillators, pressure of the system, and condensation temperature to quote a few.

Cvengroš et al. (2001) developed an experimental study for wiped film evaporator to investigate the function of roller wiper on the liquid flow regimes at different liquid loads of the evaporato s perimeter and different wiper peripheral speeds. Batistella et al. (2002) showed the potentiality of molecular distillation process for recovering vitamin E from vegetal oils. Also, the author compared the performance of the falling film and centrifugal distillers.

Lutišan et al. (2002) developed a mathematical model of evaporation from the liquid film surface at high vacuum to evaluate differences between the laminar and the turbulent regimes in the film depending on the evaporato s load and temperature.

With the aim to provide information on the choice of adequate operating conditions, Xubin et al. (2005) presented a mathematical model that describes heat and mass transfer in the film of the evaporating liquid on the heated cylinder evaporator and in the gas phase in the distillation gap by considering a binary system in the presence of an inert gas.

Sales-Cruz and Gani (2006) presented a generalized short-path evaporation model, (based on Kawala and Stepha s model), with a corresponding systematic simulation and solution strategy for various types of design/analysis problems. The author presented two case studies: the purification of a reaction mixture containing glycerol, mono-, di- and triglycerides and the recovery of a pharmaceutical product from a six-component mixture.

One of the problems arising when trying to predict the performance of those processes that involve petroleum products is the lack of experimental data and of opportunely detailed and field-proven models. This problem is emphasized for petroleum residues that cannot be removed by the atmospheric distillation (Merdrignac and Espinat, 2007).

For the sake of clarity, one of the most common petroleum residues is the so-called AR (Atmospheric Residue) pointing out the material at the bottom of the atmospheric distillation column (ASTM D 2892 method) which has an Atmospheric Equivalent Boiling Point (AEBP) higher than 673.15 K (in the following, 673.15 K+).

Petroleum residues are highly complex products which have high viscosity, significant content of heteroatoms and high molecular weights and these characteristics are directly related to the major presence of compounds with complex structures, typical of resins and asphaltenes (Merdrignac and Espinat, 2007).

There are many research papers in the literature proposing mathematical models and simulation results of the molecular distillation process for binary mixtures, in condition of low or moderate temperatures in adiabatic evaporator. Conversely, looking forward an industrial scale up of molecular distillation, other important conditions such as high process temperatures, very high vacuum pressure, and low flow rate must be analyzed. Also, the limitation of binary mixtures must be removed too.

Thus, the present research activity mainly deals with these objectives by considering the atmospheric residue Gamma + Sigma 673.15 K+ as a multicomponent mixture consisting of 6 pseudocomponents, which were characterized by availing from Petrobras simulator (PETROX) and from correlations for heavy fractions (Niederberger et al., 2005, Pedersen et al., 1984 and Poling et al., 2004). To perform the molecular distillation of the selected petroleum residue, a falling-film molecular distiller with roller wiper system and constant heated in the evaporator wall is adopted in the present work.

EXPERIMENTAL CAMPAIGN

Atmospheric Residue Gamma + Sigma 673.15 K+

A simulation of the complete fractionation of crude oil Gamma + Sigma was performed by PETROX. This simulation enabled to predict the total number of evaporated components during the conventional distillation of petroleum. The results showed a total number of 102 components which comprise from gases and vapor (water, nitrogen, methane) to components that evaporate at temperatures higher 1000 K. From the complete list, the evaporated components from 645.15 K to 1084.15 K were considered as being the constitutive of atmospheric residue. The High-Temperature Simulated Distillation (HTDS) by Gas Chromatography (GC) of atmospheric residue enabled to define the Initial Boiling Point (IBP) of most volatile component of the residue (Zuñiga et al., 2010a and Zuñiga et al., 2010b). HTDS measurements were made by the Research Center of Petrobras-Brazil (CENPES/Petrobras). The physicochemical characterization of the total components present in the residue, such as AEBP, molecular weight, specific gravity, API gravity and the percentage vaporized (%, v/v) are listed in Table 1. The list shows a total of 61 components which are grouped into 6 pseudocomponents, according to similarities of their properties and specifically the API gravity. Each group pseudocomponent is denoted by a letter: "a", "b", "c", "d", "e" and "f". The pseudocomponents present different features and different amount of components involved: the pseudocomponent "a" is the most volatile one and it consists of 5 components, whereas the pseudocomponent "f" is the heaviest one as well as the largest since it includes 19 components.

Table 1: Estimation of properties for the pseudocomponents adopted to characterize the atmospheric residue Gamma + Sigma 673.15 K+ and initial molar compositions of the inlet flow rate. $m_0 = 2.44 \times 10^{-3}$ kg s^{-1} is considered as basic value for the calculations

Pseudocomponent	Component	AEBP (K)	M_w	SG	API	% (v/v)	T_{AVBPq} (K)	SG_q	M_{wq}	x_{q0}
"a"	1	645.15	298	0.9273	21.0935	1.0418	657.7	0.9335	311	0.1312
	2	651.15	304	0.9304	20.5833	1.0620				
	3	657.15	311	0.9336	20.0668	1.0817				
	4	663.15	318	0.9366	19.5804	1.1006				
	5	670.15	325	0.9395	19.1097	1.1594				
"b"	6	675.15	332	0.9420	18.7164	1.0595	690.2	0.9474	351	0.1572
	7	680.15	338	0.9440	18.3911	0.9566				
	8	685.15	344	0.9460	18.0780	1.0093				
	9	690.15	350	0.9477	17.8160	1.0205				
	10	695.15	357	0.9492	17.5656	1.0674				
	11	700.15	363	0.9507	17.3363	1.0631				
	12	705.15	370	0.9520	17.1276	1.0586				

"c"	13	710.15	377	0.9533	16.9308	1.1041	733.7	0.9565	414	0.2177
	14	715.15	384	0.9540	16.8218	1.1127				
	15	720.15	392	0.9547	16.7152	1.1453				
	16	725.15	399	0.9553	16.6140	1.0738				
	17	730.15	406	0.9560	16.5187	1.0008				
	18	734.15	413	0.9566	16.4196	1.0302				
	19	739.15	421	0.9573	16.3190	1.0339				
	20	743.15	428	0.9579	16.2170	1.0371				
	21	748.15	435	0.9586	16.1138	1.0396				
	22	753.15	443	0.9588	16.0857	1.0412				
	23	757.15	451	0.9591	16.0416	1.0417				
"d"	24	762.15	458	0.9595	15.9794	1.0408	777.2	0.9617	484	0.1211
	25	766.15	466	0.9600	15.8993	1.0388				
	26	771.15	474	0.9606	15.8033	1.0192				
	27	776.15	483	0.9614	15.6819	1.0868				
	28	781.15	493	0.9624	15.5319	1.1531				
	29	787.15	503	0.9635	15.3636	1.1305				
	30	792.15	513	0.9647	15.1757	1.1239				

"e"		830.7	0.9764	586	0.1706	
31	798.15	523	0.9661	14.9656	1.1164	
32	803.15	533	0.9677	14.7175	1.1081	
33	809.15	544	0.9695	14.4590	1.099	
34	814.15	554	0.9712	14.1908	1.0892	
35	820.15	565	0.9730	13.9234	1.0402	
36	826.15	577	0.9749	13.6389	1.0633	
37	832.15	589	0.9770	13.3370	1.0858	
38	838.15	602	0.9790	13.0386	1.0349	
39	844.15	615	0.9810	12.7339	1.0218	
40	851.15	628	0.9834	12.3909	1.0079	
41	857.15	642	0.9858	12.0376	0.9935	
42	863.15	655	0.9882	11.6908	0.9786	

Experimental Campaign, Modeling, and Sensitivity Analysis....

"f"	869.15	669	0.9904	11.3650	0.9225	976.7	1.0157	921	0.2022
43	869.15	669	0.9904	11.3650	0.9225				
44	877.15	686	0.9931	10.9782	1.0994				
45	886.15	708	0.9962	10.5334	1.2695				
46	895.15	730	0.9992	10.1153	1.1987				
47	904.15	753	1.0021	9.7097	1.1679				
48	913.15	777	1.0049	9.3167	1.1361				
49	922.15	801	1.0075	8.9440	1.1037				
50	932.15	826	1.0101	8.5879	1.0708				
51	941.15	852	1.0126	8.2404	1.0375				
52	950.15	879	1.0151	7.9018	1.0037				
53	959.15	907	1.0175	7.5726	0.9697				
54	968.15	936	1.0197	7.2619	0.9098				
55	978.15	970	1.0225	6.8824	1.1042				
56	991.15	1012	1.0258	6.4396	1.2790				
57	1003.15	1056	1.0289	6.0283	1.1799				
58	1016.15	1102	1.0318	5.6453	1.0913				
59	1034.15	1171	1.0357	5.1164	1.4965				
60	1059.15	1275	1.0407	4.4723	1.8067				
61	1084.15	1389	1.0449	3.9195	1.5378				
Total					66.9574[b]				0.9999

[a]AEBP, M_w, SG, API and % (v/v) are the Atmospheric Equivalent Boiling Point, the molecular weight, the specific gravity (60 °F/60 °F), the API gravity and the percent vaporized of components constituted of atmospheric residue Gamma + Sigma 673.15 K+ respectively. T_{AVBPq}, SG_q, Mw_q, x_{q0} are the VABP, the specific gravity (60 °F/60 °F), the molecular weight and the mole fraction of pseudocomponents of the falling film respectively.

[b]Percentage about of the total crude oil.

Molecular Distillation

In order to fractionate the atmospheric residue, a falling film molecular distillation unit KDL 5-Mini Pilot Plant/UIC-GmbH was used. The basic design of the equipment is a short-path distillation unit, as shown inFig. 1: a vertical double-jacketed cylinder (evaporator) with a cooled and centered internal condenser, and a rotating roller wiper basket (composed of four rollers) with an external device. The equipment also has a feed device with gear pump, rotating carousels that hold discharge sample collectors for the products from the molecular distillation process (each carrousel consists of six collectors that can be positioned and moved by the operator without interrupting the distillation process), a set of vacuum pumps with an in-line low-temperature cold trap and four heating units. The roller wipers are positioned on the internal surface of the evaporator and their principal function is to homogenize the formed film and to enable easier heat transfer from the inner layers (in contact with the evaporator wall) to the film surface. The molecular distillation process occurs in steady-state, thus the atmospheric residue is continuously fed and two product flows are continuously generated: the distillate cut and the residue of molecular distillation. As soon as all temperatures (feed temperature, evaporator temperature, condenser temperature and products temperatures) and the vacuum pressure are reached, the wiper system is started. Then, the rotating gear pump feeds the sample at a constant rate onto a rotating distribution plate from a heat feed container that, for these petroleum residue, must be about 353.15 K. Centrifugal gravity forces the material distribution onto the inner surface of the evaporator in the form of a very thin film, with a thickness which will depend on the mixture's viscosity and feeding flow. Volatile components vaporize from the film and condense on the cooled inner condenser. The separation takes place in four basic stages:

- Transport of evaporated compounds from the liquid mixture to the film surface;
- Distillation of the mixture from the film surface;
- Transport of evaporated molecules through distillation gap (space between evaporator and condenser);
- And condensation of evaporated molecules.

The most volatile components which are not condensed are collected in the cold trap. Distillate cuts and residues from molecular distillation are collected separately in reservoir cylinders assembled in two carousels. The heating fluid (JULABO thermaloil) circulating through the double jacketed cylinder provides the heat for the wall evaporator. The cooling fluid (water at a constant temperature) cools the condensing cylinder. The vacuum pressure in the distillator is set by a rotary vane pump and a diffusion pump. The residence time depends of the molecular distillation conditions, especially on the evaporator temperature. It ranges between 5 min and 8 min. In this work a collecting time of 15 min was used, after that, the feed to the evaporator is interrupted to change the molecular distillation conditions. During this transition, the pressure and the condenser temperature are kept constant.

Experimental Factorial Design

In order to formulate an experimental correlation of the overall distillate flow rate Dzexp (kg s^{-1}) as a function of the more sensitive variables of the process, an experimental factorial design (EFD) is used. The EFD experiments are planned consisting of 2^3 trials plus 3 experiments on the central point to estimate the error repetition (the Pure error). So, here, 11 total experiments are made. The feed temperature T_0 (K), the feed flow rate Q_0 (L h^{-1}) and the evaporator molecular distillation temperature (TDM) (K) are the independent variables and the overall distillate flow rate is the dependent variable (response). Experiments are conducted at constant pressure (0.1 Pa) with T_0 ranging from 333.15 to 373.15 K, Q_0 ranging from 0.47 to 1.68 L h^{-1} and TDM ranging from 398.15 to 603.15 K. The most significant effects and their influences on the process response are determined by means of the software Statistica 5.0.

Table 2 shows the molecular distillation conditions used in EFD and the experimental data obtained for the atmospheric residue Gamma + Sigma 673.15 K+.

Table 2: Experimental factorial design for the prediction of the overall distillate flow rate as a function of the process variables

Residue	Gamma + Sigma 673.15 K+
P (Pa)	0.1
RA (rpm)	250
Tc (K)	363.15

Trial	T_0 (K)	TDM (K)	Q_0 (L h^{-1})	Dzexp (kg s^{-1})
1	333.15 (−1)	398.15 (−1)	0.47 (−1)	5.917 × 10^{-6}
2	373.15 (1)	398.15 (−1)	0.47 (−1)	6.300 × 10^{-6}
3	333.15 (−1)	603.15 (1)	0.47 (−1)	9.202 × 10^{-5}
4	373.15 (1)	603.15 (1)	0.47 (−1)	9.137 × 10^{-5}
5	333.15 (−1)	398.15 (−1)	1.68 (1)	6.650 × 10^{-6}
6	373.15 (1)	398.15 (−1)	1.68 (1)	4.786 × 10^{-6}
7	333.15 (−1)	603.15 (1)	1.68 (1)	2.906 × 10^{-4}
8	373.15 (1)	603.15 (1)	1.68 (1)	2.936 × 10^{-4}
9	353.15 (0)	500.65 (0)	1.08 (0)	1.088 × 10^{-4}
10	353.15 (0)	500.65 (0)	1.08 (0)	1.089 × 10^{-4}
11	353.15 (0)	500.65 (0)	1.08 (0)	1.094 × 10^{-4}

[a]Numbers in parentheses are code symbols for levels of independent variables where (−1) is the lowest level (1) is the highest level and 0 (zero) is the central point. P is the pressure of the molecular distillation process; RA is the agitation rate in the evaporator; T_c is

the condenser temperature; T_0 is the feed temperature; TDM is the evaporator temperature; Q_0 is the feed flow rate; D_{zexp} is the overall distillate flow rate obtained by experimental measurement.

The residue is identified by the IBP of the most volatile components of the mixture; values within brackets are the level of the independent variable: −1, 0, and +1 are the lowest, the central, and the highest point, respectively.

Statistic analysis by using the p-test, and the normal plot provided that, the TDM and the Q_0 are the more influence on the process than the T_0. On the other hand, F-test and analysis of variance (ANOVA) showed that the formulated model has high statistical significance, because the calculated F-value (F_{cal}) is about 255 times larger than the listed F-value (F_{list}), for 95% of confidence level (F_{cal} = 1111.94; F_{list} = 4.35). Also, the coefficient of determination of statistic model was R^2 = 0.9979 and the Pure error is 9.75 × 10⁻⁴. The EFD carried out here allowed generate an initial statistical model (1) which is reasonably reliable to suitably represent the system behavior. It allows to predict the experimental overall distillate flow rates on several molecular distillation conditions.

$$D_{zexp} = b_0 + b_1 X_1 + b_2 X_2 + b_2 X_1 X_2 \qquad (1)$$

where b_0 = 1.02 × 10⁻⁴, b_1 = 9.30 × 10⁻⁵ and b_2 = 5.00 × 10⁻⁵ are the parameter estimations coming from the regression of the experimental data; X_1 and X_2 are the independent variables described by the expressions: X_1 = [(TDM − 500.65)/102.50] and X_2 = [(Q_0 − 1.07)/0.60].

Physico-Chemical Characterization

To adjust the parameter estimations of physico-chemical properties, the analysis of viscosity, density, molecular weight, SARA separation (Brandão Pinto, 2002 and De Andrande Ferreira and Radler De Aquino Neto, 2005) (used to obtain saturated, aromatic, resins and asphaltenes fractions) and specific heat are performed on the atmospheric residue fed to the molecular distillation unit.

Also, the same analyses are carried out on the obtained products to verify the efficiency of the molecular distillation to separate the atmospheric residue. Specific information and operating conditions of these tests are reported hereinafter.

Viscosity Analysis

The viscosity measurements were performed by using a HAAKE RheoStress 6000-UTC rheometer, with plate–plate sensor system, of (LDPS/UNICAMP). At least 50 date of viscosity as a function of temperature are generated in the range of temperature 353.15–483.15 K and constant shear rate at 10 s^{-1}. Viscosity profiles against temperature allow adjusting the parameter estimation within the Amin-Beg correlation (Amin and Beg, 1993) as shown in Eq. (2):

$$Ln\eta = Ln[P_1 + P_2 \exp(-P_3(-P_4 + P_5 T_b - P_6 T_b^2))] \\ + (-P_4 + P_5 T_b - P_6 T_b^2)\left(\frac{1}{T}\right) \quad (2)$$

where η is the kinematic viscosity (m^2 s^{-1}), Tb is the 50% boiling point (K) and P_1 to P_6 are the experimental parameters whose values for Gamma + Sigma 673.15 K+ residue are: $P_1 = 1.795 \times 10^{-10}$; $P_2 = 3460.00$; $P_3 = 0.3684$; $P_4 = 1.579 \times 10^5$; $P_5 = 372.77$; $P_6 = 0.212$. This equation was used to predict viscosity-temperature behavior of the falling–film during the process simulation. Details about this adjustment have been described elsewhere (Zuñiga et al., 2010a and Zuñiga et al., 2010b).

Density Analysis

Specific gravity and density sensitivity against temperature (278.45–343.15) K of both atmospheric and molecular distillation residues are measured by means of the Pycnometer method (ASTM D 70). Density sensitivity against temperature is defined by means of d (O'Donnell, 1980):

$$\rho = \sqrt{\rho_{15.5}^2 + d(t - 15.5)} \quad (3)$$

where ρ is the density (g cm^{-3}) at temperature t (°C), $\rho_{15.5}$ is the density at standard temperature, 15.5 °C. d = −1.1 × 10^{-3} g^2 cm^{-6} °C^{-1} for the Gamma + Sigma 673.15 K+ residue.

Eq. (3) correlates density and temperature of the falling-film during the molecular distillation of the atmospheric residue. Details about the calculation of the k-value have been already described elsewhere (Zuñiga, 2009).

Molecular Weight Analysis

Molecular weight of the atmospheric residue and of the molecular distillation products are defined from VPO (Vapor Pressure Osmometry). The measurements are acquired at 328.15 K by a Knauer Vapor Pressure Osmometer, with a universal probe, where Benzyl (M = 210.23 kg kmol^{-1}) is used as standard basis and toluene is adopted as solvent. The concentration range varies from 2 to 10 g$_{solute}$ kg$_{solvent}^{-1}$ for all samples analyzed. The experimental procedure by Sabadini et al. (1997) and ASTM D 2503 method is carried out. Molecular weight measurements allow to validate parameter estimations of Eq. (4), used to predict the molecular weight of distillated pseudocomponents during the molecular distillation of the atmospheric residue (Zuñiga et al., 2008 and Zuñiga et al., 2011):

$$M_{CC} = 284.75[\exp(0.00322(t_{VABP} + 273.15))][\exp(-2.52SG)]$$
$$\times (t_{VABP} + 273.15)^{0.083} SG^{2.44} \qquad (4)$$

where M_{CC} is the molecular weight; SG is the specific gravity (1.0164) for atmospheric residues and t_{VABP} is the Volume Average Boiling Point (VABP) in °C. Table 1 summarizes the molecular weight of pseudocomponents calculated by Eq. (4).

SARA Fractionation

The results of SARA separation on atmospheric residue were used to define the value of several parameters of the Latini's equation (Poling et al., 2004):

$$\lambda_q = \frac{A_{cond}(1 - T_{VABPrq})^{0.38}}{T_{VABPrq}^{1/6}} \quad (6)$$

$$A_{cond} = \frac{A^*_{cond} T_{VABPq}^{ALFA_{cond}}}{M^{BETA_{cond}} Tc_q^{GAMMA_{cond}}} \quad (7)$$

where T_{VABPrq}, T_{VABPq}, and T_{cq} are the reduced VABP, the VABP, and the critical temperature in K of the pseudocomponent q, respectively; A_{cond}, A^*_{cond}, $ALFA_{cond}$, $BETA_{cond}$, and $GAMMA_{cond}$ are parameters whose values depend of classes of organic compounds that compose the liquid mixture (Poling et al., 2004). Details about the prediction of these parameters are described in Section 6.1.

SARA fractionation is carried out by CENPES/Petrobras, by using the technique of thin layer chromatography (TLC) with field ionization detector (FID). The instrument and its operation are described in ASTM D 4124 method.

Latini's equation is used to predict the thermal conductivity of pseudocomponents that compose the inlet flow to distiller and the evaporated pseudocomponents during the molecular distillation process. Rowley's method (Poling et al., 2004), which defines the thermal viscosity of a liquid mixture, is used to calculate the thermal conductivity of the atmospheric residue fed to the molecular distillation unit:

$$\lambda_{mis} = \left(\sum_{q=1}^{N} w_q \lambda_q^{-2} \right)^{-1/2} \quad (5)$$

where λ_{mis} is the thermal conductivity of falling film (W m K^{-1}); w_q is the weight fraction of pseudocomponent q; and λ_q is the thermal conductivity of the pure pseudocomponent q (W m K^{-1}).

Specific Heat Measurements

By using the Differential Scanning Calorimeter (DSC), the specific heat of the atmospheric residue Gamma + Sigma 673.15 K+ is measured in the range 353.15–573.15 K. The experimental procedure is the one defined by ASTM E 1269 method. This analysis allows to adjust the parameter estimations for calculating the specific heat by means of a function of temperature of the falling-film during the molecular distillation process of the residue. More details of the experimental conditions are reported by Ballesteros et al. (2009). The correlation is as follows (Perry and Green, 1999):

$$CP_{mis} = A_{CP}(SG_{mis})^{-0.5} + B_{CP}(t - 15.5) \tag{8}$$

where CP_{mis} (cal g^{-1} °C^{-1}), and SG_{mis} are the specific heat and the specific gravity at 288.15 K of the falling film, respectively, t is the temperature of the falling-film in (°C). $A_{CP} = 0.32389$ and $B_{CP} = 0.00091$ are the regression parameters to fit experimental data of specific heat for the residue Gamma + Sigma 673.15 K+.

WORKING EQUATIONS

A series of assumptions are adopted to develop the generalized two-dimensional steady-state mathematical model of the molecular distillation process:

- The evaporator has a cylindrical shape and it is internally heated by a stream at the constant temperature T_w.
- The condensing cylinder is cooled by a stream at the constant temperature T_c.
- There is a falling-film that flows down while it is partially vaporized.
- The falling-film flows at laminar conditions along the vertical axis and no waves appear on the falling-film surface.
- Liquid films on both the evaporation and the condensation

walls have a negligible thickness with respect to the unit diameter.
- The mass, heat and momentum balances were developed in cylindrical coordinates.
- Liquids have Newtonian behavior.
- The portion of liquid that evaporates is highly viscous; hence, Reynolds numbers are small.

Re-evaporation and splashing phenomena are both neglected.

No axial diffusion and radial flow phenomena are considered.

The mathematical model is based on the above assumptions, on the original study by Kawala and Stephan (1989), and on momentum, energy and mass balances for both the evaporation and the condensation films.

Velocity Distribution of Liquid Film

Fig. 2 reports the velocity profile in the liquid film. The film velocity (W_z) can be expressed as:

$$W_z(r, z) = \frac{g\rho_{mis}}{\eta_{mis}} S^2 \left[\frac{R-r}{S} - \frac{1}{2}\left(\frac{R-r}{S}\right)^2 \right] \tag{9}$$

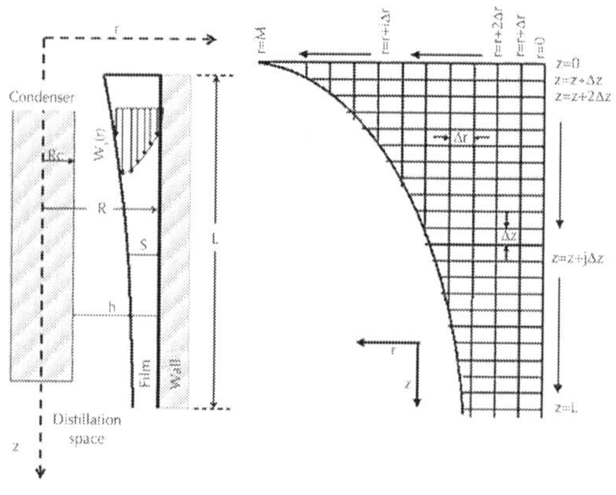

Figure 2: Velocity profile, film shape on the short-path evaporator and discretization scheme in the film defined by using the finite-difference

method. h is the distance between the evaporator and the condenser; L is the film length; M is the maximum number of discrete points in the radial coordinate; r is the radial coordinate; R is the internal radius of evaporator; Rc is the outer radius of condenser; S is the film thickness; i and j are the positive increments in the radial and axial coordinate respectively; z is the axial coordinate; Δr and Δz are the distance between two discrete points on radial and axial coordinate respectively.

It can be observed that the term [(R − r)/S] is modified by considering the concave configuration of the short-path distillation unit used in the present research activity, where the condenser is wholly surrounded by the evaporator. Thus, the evaporation occurs on the inner surface of the outer cylinder and the condensation at the outer surface of the inner cylinder (Xubin et al., 2005).

Thickness of the Evaporating Film

The thickness film is essential to evaluate the velocity profile. This varies with the film length because of the axial and radial gradients of the temperature within the liquid film. In addition, variations in the evaporation rate of molecules are generated on the film surface. The thickness of the falling film is expressed as follows:

$$S = \left[3\eta_{mis} \left(\frac{m_0}{2\pi g R \rho_{mis}^2} - \frac{1}{g\rho_{mis}^2} \int_{z_0}^{z} G_g \, dz \right) \right]^{\frac{1}{3}} \quad (10)$$

Surface Evaporation Rate under Vacuum

The effective rate G_q of evaporation of the pseudocomponent q is calculated through a modified Langmuir–Knudsen equation: the modification is based on the assumption that the anisotropic properties of the vaporized molecules fade if the number of collisions is larger than two (Kawala and Stephan, 1989). Eq.(11) contains the factor (P/P_{ref}), introduced by Sales-Cruz and Gani (2006) to correct the vacuum pressure;P is the system pressure and P_{ref} is the effect of pressure system which is related with the increase the system pressure due the contribution of two classes of molecules: the residual gas and the distillate vapor. These pressures are referred to as residual pressure and saturation pressure, respectively. Hickman (1943) affirmed that the rate of collection of distillate approaches the rate of evaporation only at a zero residual pressure and a short unobstructed path. With wide gaps and substantial residual pressures, few molecules will reach the condenser without collision and many of them will go back to the falling film. In the present activity the value of P_{ref} is obtained by means of specific tests and of an ad hoc software (DESTMOL-P) (Zuñiga, 2009) just developed for this case study.

$$G_q = x_{qS} P_q^{vap} \left(\frac{M_q}{2\pi R_g T_S} \right)^{1/2} \left(\frac{P}{P_{ref}} \right) [1 - (1-F)(1 - e^{-h/k\beta})^n] \tag{11}$$

is a surface ratio and k is the degree of anisotropy of the vapor phase in the distillation space between the evaporator and the condenser. Where $F = A_k/(A_k + A_v)$ is evaluated by the expression $\log k = 0.2F + 1.38(F + 0.1)^4$. In the case of distillation of a multicomponent mixture, the evaporation rate is calculated as follows:

$$G_g = \sum_{q=1}^{N} x_q G_q \tag{12}$$

Temperature Profile in the Liquid Film

In the case of a heated evaporator, there is a heat exchange between the inner surface of evaporator and the falling film, which improves the enthalpy of vaporization of the liquid. The evaporation takes place at the free surface of the liquid and the heat required is supplied from the deeper layers by conduction and forced convection phenomena. This produces a radial temperature gradient in the liquid by perturbing the evaporation rate. The temperature in the liquid film is defined by Eqs. (13)–(16) (Kawala and Stephan, 1989).

$$W_z \frac{\partial T}{\partial z} = \alpha_{mis} \left[\frac{1}{r} \frac{\partial T}{\partial r} + \frac{\partial^2 T}{\partial r^2} \right] \quad (13)$$

The initial and boundary conditions are:

$$T = T_0 \quad \text{for} \quad z = 0 \quad \text{and} \quad (R - S) \leq r \leq R \quad (14)$$

$$T = T_w \quad \text{for} \quad r = R \quad \text{and} \quad 0 \leq z \leq L \quad \text{(Heated evaporator)} \quad (15)$$

$$\frac{\partial T}{\partial r} = -\frac{\sum_{q=1}^{N}(G_q \Delta H_q^{vap})}{\lambda_{mis}} \quad \text{for} \quad r = (R - S) \quad \text{and} \quad 0 \leq z \leq L \quad (16)$$

Eq. (14) defines the film temperature at the evaporator inlet. T_0 is the inlet feed temperature. Eq. (15) represents the film temperature profile on the wall of the distillation unit. T_w is the heating fluid temperature. Eq. (16) represents the variation of the film temperature on the surface, which depends on the evaporation dynamics.

Concentration Profile in the Liquid Film

The concentration of pseudocomponents in the liquid film varies along axial and radial directions according to the value of the evaporation rate on the film surface. The concentration in the liquid layer for a multicomponent mixture can be expressed by the following equations:

$$W_z \frac{\partial C_q}{\partial z} = D_{q\_mis} \left[\frac{1}{r} \frac{\partial C_q}{\partial r} + \frac{\partial^2 C_q}{\partial r^2} \right], \quad q = 1, 2, \ldots, N \quad (17)$$

The initial and boundary conditions are (Batistella and Wolf Maciel, 1996):

$$C_q = C_q 0 \quad \text{for} \quad z = 0 \quad \text{and} \quad (R - S) \leq r \leq R \quad (18)$$

$$\frac{\partial C_q}{\partial r} = 0 \quad \text{for} \quad r = R \quad \text{and} \quad (0 \leq z \leq L) \quad (19)$$

$$\frac{\partial C_q}{\partial r} = -\frac{G_q - C_q M_q \left(\sum_{\upsilon=1}^{N} G_\upsilon/M_\upsilon \atop \upsilon \neq q \right)}{\rho_{mis} D_{q\_mis}}$$

$$\text{for } r = (R - S) \text{ and } 0 \leq z \leq L \quad (20)$$

D_{q-mis} (m² s⁻¹) is the diffusion coefficient of the pseudocomponent q in the liquid mixture, which is determined by using the Wilke–Chang estimation method (Poling et al., 2004). The initial concentration of the pseudocomponent q of the falling film in the evaporator inlet C_{q0} (kmol m⁻³) is considered equivalent to the initial composition of the feed to the evaporator, as shown in Table 1. The boundary condition of Eq. (19) defines the concentration of the pseudocomponent q in the layer of the film beside the wall evaporator. This composition varies along the axial direction of the film. Eq. (20) defines the concentration of the pseudocomponent q on the film surface. This concentration is not constant but varies along the film, as a result of the temperature variations along the flow path on z direction.

Overall Distillate Mass Flow Rate

By definition, the overall distillate mass flow rate (D_z) is the sum of the local distillate rates, therefore, it varies along the axial direction (Kawala and Stephan, 1989). It is expressed as:

$$D_z = \sum_{z=0}^{L} G_{gz} A_z$$

$$(21)$$

where A_z represents the surface area element at the distance z from the inlet, which is evaluated as $A_z = 2\varpi(R - S)z$ for concave configuration of the short-path distillation unit.

SOLUTION STRATEGY

The partial differential equations (13)–(20) are discretized in finite differences form by using a central finite-difference method and, specifically, the Crank-Nicolson method is adopted (Carnahan et al., 1969). The resulting Algebraic Equations (AEs) system is solved using Gaussian Elimination through the subroutine Dgefa in Fortran 90. The film thickness is discretized on 10 intervals (11 discrete points), while the film length is discretized on 300 intervals (301 discrete points). The scheme discretization is shown in Fig. 2. Through such a discretization, a good simulation performance was observed. As a result, a system of 10 AEs of temperature and 66 AEs of concentration (11 for each of 6 considered pseudocomponents), which comprise 76 AEs for each axial position of evaporator, was obtained. The total AEs system is 22800, which define the temperature profile and the concentration profiles of pseudocomponents in the liquid film. The software DESTMOL-P in Fortran 90 was developed to solve the overall system and to generate the profiles of the process variable in the film. The calculation procedure to determine the process variables is described in Fig. 3.

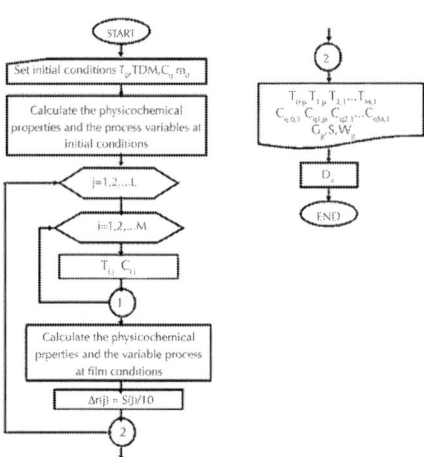

Figure 3: Flow chart to determine the variables of molecular distillation process of atmospheric residue of Gamma + Sigma crude oil.

CASE STUDY: MOLECULAR DISTILLATION OF AN ATMOSPHERIC PETROLEUM RESIDUE

The distillation of residue Gamma + Sigma 673.15 K+ is considered as case study. The residue is divided into six pseudocomponents and each of them is characterized by using PETROX and some specific correlations for the heavy fractions [21] as show in Table 1. The initial molar concentration of pseudocomponents in the inlet of the evaporator (x_{q0}) is also reported in Table 1. Eqs. (2)– (5) and (8) are used to evaluate the viscosity and density of the, falling film, the molecular weight of pseudocomponents, the thermal conductivity, and the specific heat of the falling film during the molecular distillation process, respectively. Also, physico-chemical properties such as vapor pressure, diffusion coefficient, and evaporation enthalpy are expressed in terms of VABP of fraction, molecular weight, and specific gravity.

Operating Conditions

The unit for the molecular distillation adopted in the present research has the following geometrical dimensions:

– Evaporator length:	0.31 m
– Evaporator internal radius:	0.0425 m
– Distillation gap:	0.0221 m
– Mean path of vapor molecules:	0.03 m (this value is determined from several tests by using the DESTMOL-P)
– Vacuum pressure:	0.1 Pa
– Pressure system effect:	0.559 Pa (which is defined according to the procedure of item 6.1)

– Condenser temperature Tc:	363.15 K
– Feed temperature T_0:	353.15 K
– Molecular distillation temperature TDM:	533.15 K (this temperature is equivalent to the wall temperature Tw of the evaporator)
– Inlet feed flow rate m_0:	2.44 × 10⁻³ kg s⁻¹

Accordingly, a distillate mass flow rate of 9.39 × 10⁻⁵ kg s⁻¹ and a molecular distillation residue flow rate of 2.35 × 10⁻³ kg s⁻¹ are obtained. The average composition of pseudocomponents in the distillate and the residue flow rates are summarized in Table 3.

Table 3: Average composition of pseudocomponents in the product flow rates at case study condition

Distillate flow rate (kg s⁻¹)	9.39 × 10⁻⁵
Residue flow rate (kg s⁻¹)	2.35 × 10⁻³

Pseudocomponent	xq	
	Distillate	Residue
"a"	0.7008	0.1195
"b"	0.2553	0.1515
"c"	0.0419	0.2218
"d"	0.0020	0.1242
"e"	0.0001	0.1751
"f"	0.0000	0.2079
Total	1.0000	1.0000

[a]x_q is the composition in mole fraction of pseudocomponent.

RESULTS AND DISCUSSION

Prediction of the Pressure System Effect and Estimation of the Thermal Conductivity Parameters

The values of the pressure system effect Pref and the thermal conductivity parameters A^*_{cond}, $ALFA_{cond}$, $BETA_{cond}$ and $GAMMA_{cond}$, are calculated by using several numerical tools and methods such as the observers of the statistic analysis and robust optimization algorithm. Specifically, the LCONF belonging to the IMSL numerical library of Fortran 90 has been adopted. The methodology adopted could be described on the follow step:

Step 1. Initial statistic analysis. Using an experimental fractional factorial design of resolution five (Barros et al., 2003) (EFFD), 2_V^{5-1}, the influence of pressure system effect and the thermal conductivity parameters on the overall distillate flow rate are evaluated. The sensitivity analysis is performed for a specific operating condition (TDM = 533.15 K, m_0 = 2.44 × 10^{-3} kg s^{-1}, T_0 = 353.15 K, T_c = 363.15 K and P = 0.1 Pa) and several values of the sensitive variable Dz (kg · s^{-1}) are generated for different conditions of P_{ref}, A^*_{cond}, $ALFA_{cond}$, $BETA_{cond}$ and $GAMMA_{cond}$. The independent variables are in the range of 0.1–1000.90 Pa for P_{ref}, 0.00319–0.562 for A^*_{cond}, 0.00–1.20 for $ALFA_{cond}$, and 0.5–1.0 for $BETA_{cond}$. A total 16 simulations are run by DESTMOL-P. The sensitivity matrix is summarized in Table 4. It is worth mentioning that, as the design results are obtained from the simulator, replicated trials do not necessary, thus, Pure error is not determined. Statistic analysis (p-test) confirms that neither the parameter $BETA_{cond}$ nor $GAMMA_{cond}$ influence the sensitive variable Dz.

Table 4: Fractional factorial of resolution five design used to evaluate the influence of the pressure system effect and the thermal conductivity parameters on the overall distillate flow rate

Trial	Independent variables				Response variables	
	A^*_{cond}	$ALFA_{cond}$	$BETA_{cond}$	$GAMMA_{cond}$	P_{ref} (Pa)	D_z (kg s^{-1})
1	0.00319 (−1)	0.00 (−1)	0.50 (−1)	−0.167 (−1)	1000.90 (1)	5.587 × 10^{-10}
2	0.562 (1)	0.00 (−1)	0.50 (−1)	−0.167 (−1)	0.10 (−1)	2.699 × 10^{-5}
3	0.00319 (−1)	1.20 (1)	0.50 (−1)	−0.167 (−1)	0.10 (−1)	1.069 × 10^{-3}
4	0.562 (1)	1.20 (1)	0.50 (−1)	−0.167 (−1)	1000.90 (1)	2.711 × 10^{-5}
5	0.00319 (−1)	0.00 (−1)	1.00 (1)	−0.167 (−1)	0.10 (−1)	3.861 × 10^{-6}
6	0.562 (1)	0.00 (−1)	1.00 (1)	−0.167 (−1)	1000.90 (1)	5.588 × 10^{-10}
7	0.00319 (−1)	1.20 (1)	1.00 (1)	−0.167 (−1)	1000.90 (1)	1.867 × 10^{-9}
8	0.562 (1)	1.20 (1)	1.00 (1)	−0.167 (−1)	0.10 (−1)	4.788 × 10^{-3}
9	0.00319 (−1)	0.00 (−1)	0.50 (−1)	0.167 (1)	0.10 (−1)	3.876 × 10^{-6}
10	0.562 (1)	0.00 (−1)	0.50 (−1)	0.167 (1)	1000.90 (1)	5.672 × 10^{-10}
11	0.00319 (−1)	1.20 (1)	0.50 (−1)	0.167 (1)	1000.90 (1)	6.328 × 10^{-9}
12	0.562 (1)	1.20 (1)	0.50 (−1)	0.167 (1)	0.10 (−1)	4.504 × 10^{-3}
13	0.00319 (−1)	0.00 (−1)	1.00 (1)	0.167 (1)	1000.90 (1)	5.587 × 10^{-10}
14	0.562 (1)	0.00 (−1)	1.00 (1)	0.167 (1)	0.10 (−1)	4.005 × 10^{-6}
15	0.00319 (−1)	1.20 (1)	1.00 (1)	0.167 (1)	0.10 (−1)	4.635 × 10^{-6}
16	0.562 (1)	1.20 (1)	1.00 (1)	0.167 (1)	1000.90 (1)	1.489 × 10^{-7}

[a]Numbers in parentheses are code symbols for levels of independent variables where (−1) is the lowest level and (1) is the highest level. A^*_{cond}, ALFAcond, BETAcond, GAMMAcond are the thermal conductivity parameters; P_{ref} is the pressure system effect; D_z is the overall distillate flow rate.

According to Table 10-4 reported by Poling et al. (2004) and considering the composition of Gamma + Sigma 673.15 K+ here selected defined by SARA fractionation as well (17% of saturated, 41% of total aromatic, 29% of resins and 13% of asphaltenes), the values of parameters are: BETAcond = 0.75 andGAMMAcond = 0.1336.

A correlation of D_z (kg s^{-1}) against Pref, A^*_{cond}, and ALFA$_{cond}$ which significantly influence the response variable, is formulated by availing from Statistica 5.0. As the model presents a correlation indexR^2 = 0.9839 and the standard deviation is particularly small 6.07 × 10^{-7}, it is considered a reliable model:

$$D_z = b_0 + b_1 X_1 + b_2 X_2 - b_3 X_3 + b_{12} X_1 X_2 - b_{13} X_1 X_3 - b_{23} X_2 X_3 - b_{123} X_1 X_2 X_3 \qquad (22)$$

where the regression parameters are b_0 = 6.52 × 10^{-4}, b_1 = 5.17 × 10^{-4}, b_2 = 6.47 × 10^{-4}, b_3 = 6.49 × 10^{-4}, b_{12} = 5.14 × 10^{-4}, b_{13} = 5.13 × 10^{-4}, b_{23} = 6.44 × 10^{-4} and b_{123} = 5.10 × 10^{-4}. X_1, X_2 and X_3 are the independent variables described by the expressions: X_1 = (A^*_{cond} − 0.2826)/0.2794; X_2 = (ALFA$_{cond}$ − 0.60)/0.60 and X_3 = (P_{ref} − 500.5)/500.4.

Step 2. Regression. In order to set the values of the most significant parameters, Eq. (22) is optimized by using the subroutine LCONF which minimizes a general objective function subject to linear equality/inequality constraints (The routine is based on M.J.D. Powell's TOLMIN, which solves linearly constrained optimization problems). There are many well-known and effective criteria to handle the problem of the optimal design of experiments as described in the past and recent literature (i.e.Franceschini and Macchietto, 2008). For example, Buzzi-Ferraris and Manenti (2009) recently listed different criteria and discussed their strengths and weaknesses; some of them are: the Box criterion, which tries to minimize the confidence volume of parameters (Box and Lucas, 1959 and Box, 1971); Hosten criterion, which tries to minimize the maximum diameter of the confidence volume of parameters (Hosten, 1974); Smith (1918) criterion, which minimizes the maximum variance for the model prevision. Other effective criteria minimize the maximum distance among the existing points either in the principal axes space. For the case of our interest, the EFD method is enough to provide a reasonable design of experiment for model validation purposes.

Eq. (23) describes the optimization problem where the constraints of the conductivity parameters are those ones defined by Poling (Table 10-4 in Poling et al., 2004) and the constraints of P_{ref} are dictated by operating conditions of the molecular distillation. The regression allows to define a set of values of conductivity parameters and P_{ref}, that generated a minimum difference between the experimental (D_{zexp}) and predicted (D_{zcal}) overall distillate flow rate. This set of values are reliable for the molecular distillation process of the atmospheric residue. To extend the active domain of this comparison, several D_{zexp} (kg s^{-1}) values are generated by Eq. (1) as additional experimental points.

It is observed that ALFA$_{cond}$ and P_{ref} are constant for the selected set of operating conditions: ALFA$_{cond}$ = 1.080 and P_{ref} = 0.559 Pa. Conversely, the A^*_{cond} values are not constant with the operating conditions and it is necessary to formulate a specific sensitivity correlation.

$$\min f(A^*_{cond}, ALFA_{cond}, P_{ref}) = D_z - 6.52 \times 10^{-4} - 5.17 \times 10^{-4} A^*_{cond} - 6.47 \times 10^{-4} ALFA_{cond} + 6.49 \times 10^{-4} P_{ref}$$
$$- 5.14 \times 10^{-4} A^*_{cond} ALFA_{cond} + 5.13 \times 10^{-4} A^*_{cond} P_{ref} + 6.44 \times 10^{-4} ALFA_{cond} P_{ref} + 5.10 \times 10^{-4} A^*_{cond} ALFA_{cond} P_{ref}$$

s.t. :

$$A^*_{cond} \leq 0.562; ALFA_{cond} \leq 1.20; P_{ref} \leq 1000.90$$
$$0.000319 \leq A^*_{cond} \leq 0.562; 0 \leq ALFA_{cond} \leq 1.20; 0.1 \leq P_{ref} \leq 1000.90 \quad (23)$$

Step 3. Final statistic analysis. In this step a correlation A^*_{cond} as a function of the TDM and Q_0 is formulated from statistic analysis, where through a Central Composite Rotate Design (CCRD), with a central point a matrix of 9 total trials is generated. These trials are conducted by using the DESTMOL-P at constant pressure (0.1

Pa), for an assigned set of operating conditions: T_0 = 353.15 K, T_c = 363.15 K, Q_0 is in the range 0.47–1.68 L h^{-1}, and TDM is in the range 460.15–622.15 K. According to ANOVA (R^2 = 0.9998) the resulting model is reasonably reliable:

$$A^*_{cond} = b_0 + b_1 X_1 + b_2 X_2 + b_{22} X_2^2 + b_{12} X_1 X_2 \qquad (24)$$

where b_0 = 0.2091; b_1 = 0.0288; b_2 = 0.2288; b_{22} = 0.0630 and b_{12} = 0.0264. The independent variables are described by the expressions X_1 = [(TDM − 540.65)/57.5] X_2 = [(Q_0 − 1.07)/0.60].

Thus, all parameters of the thermal conductivity equation and the pressure system effect are used in the mathematical modeling to predict the profiles of the process variables during the molecular distillation process for the case study.

Velocity Distribution in Film Surface

Fig. 4 shows the trend of the velocity of the liquid film at the vapor–liquid surface against the axial coordinate of the evaporator (film length). It is possible to observe that the velocity gradually increases along the evaporator. It reaches a minimum of 6.9 × 10^{-3} m s^{-1} at the inlet region and a maximum of 9.3 × 10^{-2} m s^{-1} at the outlet region. Such a behavior is due to the fact that this variable is strongly influenced by the evaporator temperature. Since the temperatures are low at the inlet flow rate, the liquid film is characterized by a high viscosity and hence the velocity decreases. On the other hand, the temperatures are high at the outlet flow rate and therefore a low viscosity characterize the film with a consequent increase in velocity.

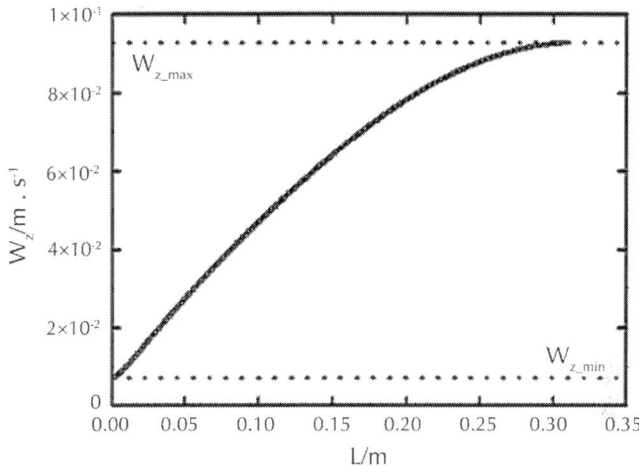

Figure. 4: Velocity distribution of the liquid film on the surface as a function of the film length for the case study conditions. $W_{z\_min} = 6.9 \times 10^{-3}$ m s^{-1} is the minimum velocity value at the inlet region of evaporator; $W_{z\_max} = 9.3 \times 10^{-2}$ m s^{-1} is the maximum velocity value at the outlet region of evaporator.

Temperature Distribution in Film Surface

Fig. 5 shows the temperature in the film surface TS (K) along the axial coordinate (film length) for $m_0 = 2.4 \times 10^{-3}$ kg s^{-1} at three different temperatures of molecular distillation, TDM = 483.15 K, TDM = 533.15 K and TDM = 598.15 K. It is observed that the T_S progressively increases throughout the film length and it reaches a maximum in correspondence with the output region of evaporator. Also, T_S increases with the molecular distillation temperature, especially from the center region for output region of evaporator, where the evaporation of the most volatile pseudocomponents is the highest one. The largest value in the film temperature difference is equal to 77 K and it is observed in correspondence with the highest distillation temperature (598.15 K).

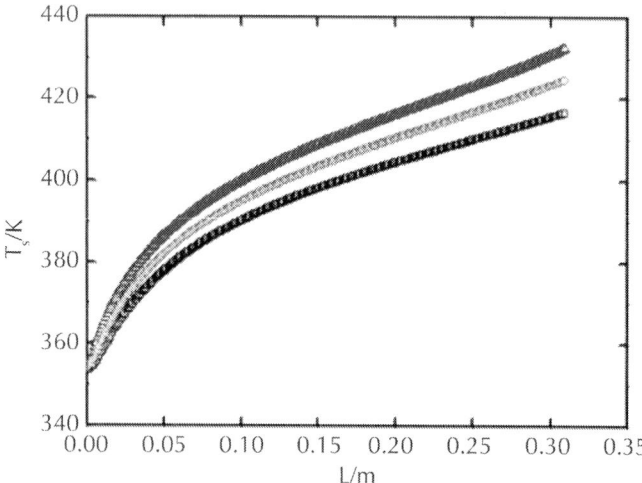

Figure. 5: Temperature profile in the film surface against the film length for a constant feed mass flow rate. $m_0 = 2.4 \times 10^{-3}$ kg s^{-1}. T_s at TDM = 483.15 K; ○T_s at TDM = 533.15 K; ΔT_s at TDM = 598.15 K.

Distribution of the Thickness Film

The thickness film profile along the evaporator for the case study is shows in Fig. 6. It decreases continuously through the evaporator, from a maximum of 2.2×10^{-3} m in the inlet region of the film, down to a minimum of 1.7×10^{-3} m in correspondence with the end region of the film. Such a decrease of the thickness film could be explained by the depletion of the most volatile pseudocomponents and the increase of the temperature in the liquid film, as well caused by the heat exchange between the inner surface of evaporator and the falling film.

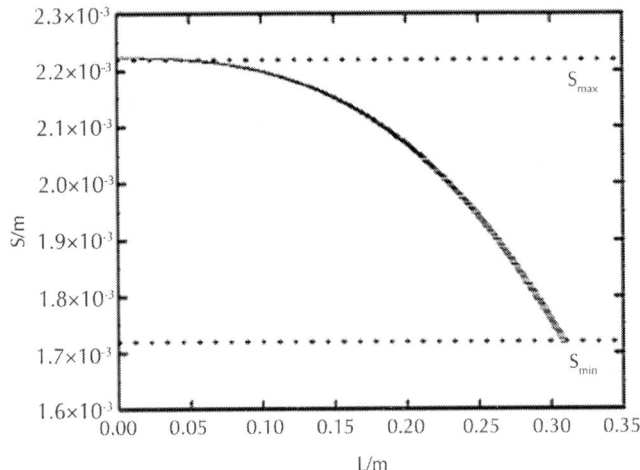

Figure. 6: Thickness film profile as a function of film length for the case study. $S_{min} = 1.7 \times 10^{-3}$ m is the minimum thickness film in the inlet zone of the film; $S_{max} = 2.2 \times 10^{-3}$ m is the maximum thickness film in the outlet zone of the film.

Concentration Distribution

The concentration distribution of six pseudocomponents at the surface of the evaporating film, as a function of film length for the conditions of case study is shown in Fig. 7. As observed, the pseudocomponents "a" and "b" were evaporated at molecular distillation conditions of case study. A gradually decrease of concentration of these pseudocomponents was obtained in the liquid film. These pseudocomponents constitute the distillate flow rate. The "c", "d", "e", and "f" pseudocomponents were not evaporated, and their concentration increase in the liquid film testifies it. This behavior indicates that this pseudocomponents remain unchanged during the process and that formed the residue flow rate of molecular distillation. Fig. 8 shows the concentration profile at the film surface of the "a" pseudocomponent (the most volatile) at three different temperatures of molecular distillation. The rapid decrease of the concentration in the liquid film shows that the liquid phase is getting poorer in this pseudocomponent, which is faster when the temperature of distillation rises. For "a" pseudocomponent the maximum concentration gradient is observed for the highest temperature. For instance, for a feed mass flow rate ofmo = 2.4×10^{-3} kg s^{-1} and TDM = 598.15 K, the concentration of "a" decreases from 0.1312 to 0.089, which corresponds to −39%. On another hand, the concentration profile of "f" pseudocomponent presented an opposite behavior when compared with the "a" pseudocomponent. As is shown in Fig. 9, the concentration increases gradually along the liquid film. Also, the maximum concentration gradient is presented for the highest temperature and the lowest feed flow rate.

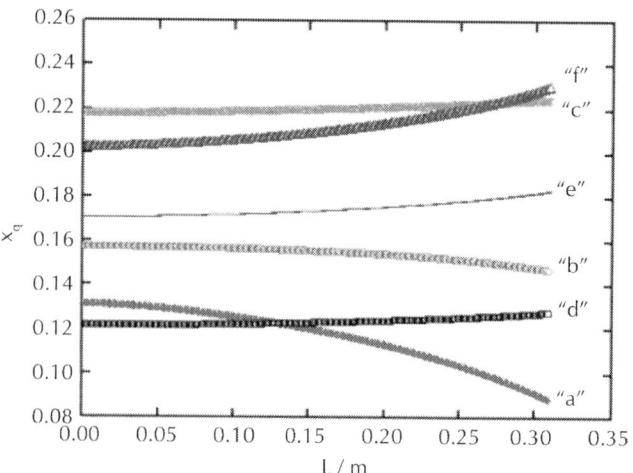

Figure. 7: Concentration profiles of pseudocomponents in the film surface x_q (mole fraction) against the film length for the molecular distillation conditions of case study. + "a" pseudocomponent; ○ "b" pseudocomponent; × "c" pseudocomponent; "d" pseudocomponent;_− "e" pseudocomponent; Δ "f" pseudocomponent.

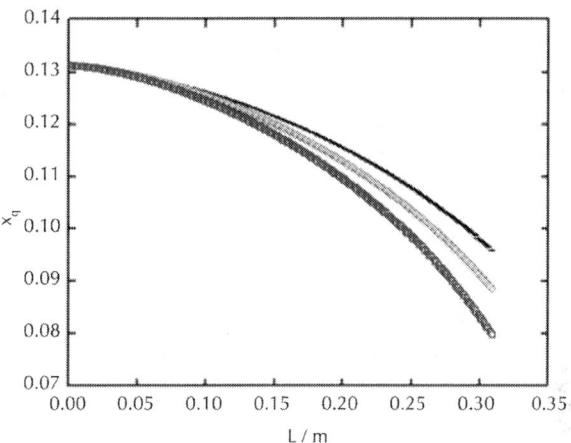

Figure. 8: Concentration profile of "a" pseudocomponent (the most volatile) in the film surface as a function of film length for a constant feed mass flow rate. $m_0 = 2.4 \times 10^{-3}$ kg s^{-1}; – TDM = 483.15 K; ○ TDM = 533.15 K; TDM = 598.15 K.

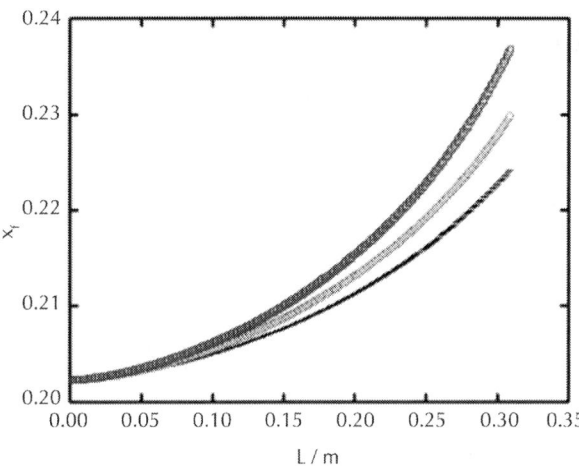

Figure. 9: Concentration profile of "f" pseudocomponent (the heaviest) in the film surface as a function of film length for a constant feed mass flow rate. $m_0 = 2.4 \times 10^{-3}$ kg s^{-1}; – TDM = 483.15 K; ○ TDM = 533.15 K; TDM = 598.15 K.

Evaporation Rate Distribution

Fig. 10 shows the evaporation profile in the film surface as a function of molecular distillation temperature, along the evaporator. The evaporation rate is strongly influenced by the molecular distillation temperature, thus, it is observed an increase in the evaporation rate of 300 times with respect the initial value when the distillation temperature increases a fixed feed mass flow rate. The rapid increase the evaporation rate in the outlet region of evaporator is due to the temperature increase of the liquid film because of the depletion of the most volatile pseudocomponents.

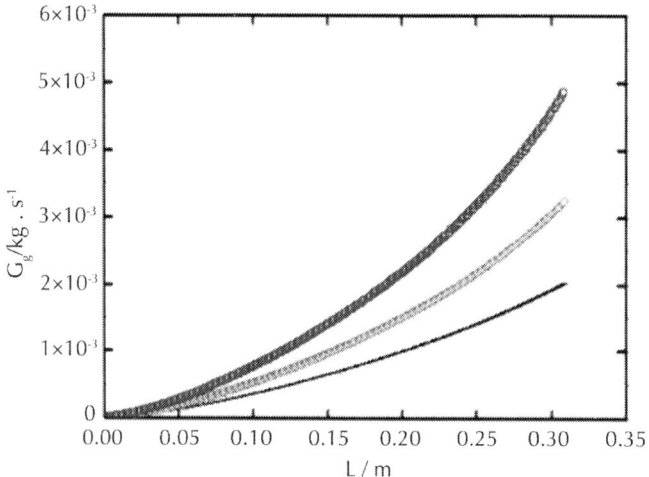

Figure 10: Evaporation rate G_g (kg s^{-1}) profile in the film surface as a function of molecular distillation temperature along the evaporator. $m_0 = 2.4 \times 10^{-3}$ kg s^{-1}; − TDM = 483.15 K; ∘ TDM = 533.15 K; □ TDM = 598.15 K.

Overall Distillate Flow Rate Distribution

The overall distillate flow rate (D_z) profiles as a function of molecular distillation temperature (TDM) and the feed flow rate (m_0), along the evaporator are shown in Fig. 11 and Fig. 12 respectively. D_z increases when the TDM and the m_0 increase. For example, to TDM = 598.15 K (the highest temperature) and for the operational conditions of case study, the D_z is highly influenced by the m_0. Moreover, the highest amounts of distillate are generated from the middle region of evaporator, where the evaporation of most volatile compounds is presented.

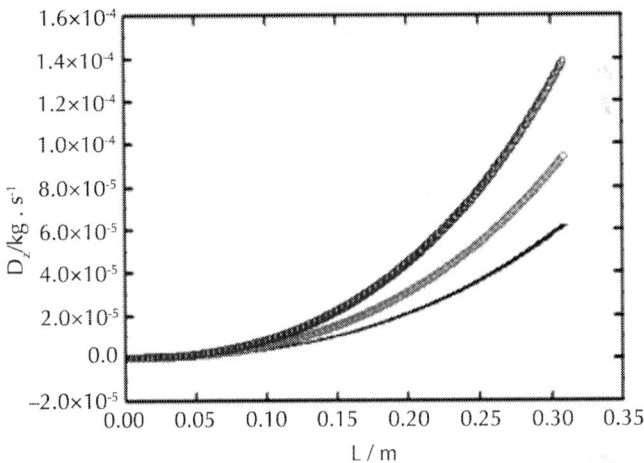

Figure 11: Overall distillate flow rate D_z (kg s^{-1}) profile as a function of molecular distillation temperature along the evaporator. m_0 = 2.4 × 10^{-3} kg s^{-1}; – TDM = 483.15 K; ∘ TDM = 533.15 K; □ TDM = 598.15 K.

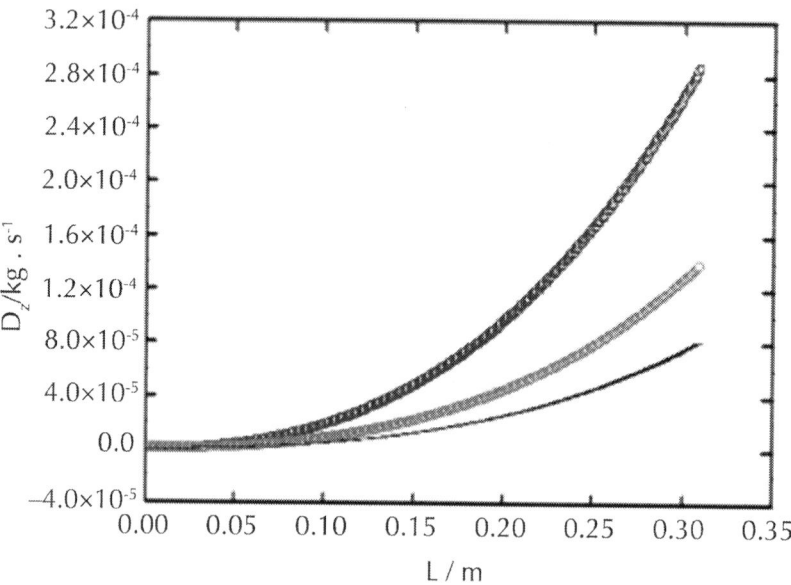

Figure 12: Overall distillate flow rate D_z (kg s^{-1}) profile as a function of feed mass flow rate along the evaporator. TDM = 598.15 K; $-m_0$ = 1.5 × 10^{-3} kg s^{-1}; ○ m_0 = 2.4 × 10^{-3} kg s^{-1}; □ m_0 = 5.5 × 10^{-3} kg s^{-1}.

The dependency of D_z with the TDM and the m_0 was demonstrated by Kawala and Stephan (1989) during the molecular distillation of a binary mixture in an adiabatic distillator. Table 5 summarized the values of D_{zcal} predicted by using the DESTMOL-P software, for different TDM and m_0. These predicted results are compared with the experimental measurements D_{zexp} obtained during the molecular distillation of residue Gamma + Sigma 673.15 K+. Some experimental values were also determined by the experimental equation (Eq. (1)) defined by the statistic analysis. In Table 5 the Average Absolute Deviation (AAD) between the experimental and the predicted overall distillate flow rate is calculated using the following equation:

$$AAD = \left(\frac{1}{n}\right) \sum_{i=1}^{n} \frac{|D_{zcal} - D_{zexp}|}{D_{zexp}} \times 100 \tag{25}$$

Table 5: Overall distillate flow rate predicted and experimental at different molecular distillation temperatures and feed flow rate.[a]

TDM (K)	m_0 (kg s^{-1})	D_{zexp} (kg s^{-1})	D_{zcal} (kg s^{-1})	AAD %
483.15	2.4 × 10^{-3}	6.35 × 10^{-5}	6.18 × 10^{-5}	2.69
493.15		6.70 × 10^{-5}	6.82 × 10^{-5}	2.46
503.15		7.64 × 10^{-5}	7.46 × 10^{-5}	2.29
513.15		8.28 × 10^{-5}	8.10 × 10^{-5}	2.15
533.15		9.57 × 10^{-5}	9.39 × 10^{-5}	1.88
543.15		1.02 × 10^{-4}	1.00 × 10^{-4}	1.66
553.15		1.09 × 10^{-4}	1.07 × 10^{-4}	1.56
563.15		1.15 × 10^{-4}	1.14 × 10^{-4}	1.22
573.15		1.21 × 10^{-4}	1.20 × 10^{-4}	0.74
598.15		1.38 × 10^{-4}	1.38 × 10^{-4}	0.65
533.15	1.54 × 10^{-3}	6.53 × 10^{-5}	6.53 × 10^{-5}	0.02
	1.99 × 10^{-3}	8.05 × 10^{-5}	7.91 × 10^{-5}	1.74
	2.90 × 10^{-3}	1.11 × 10^{-4}	1.09 × 10^{-4}	1.62
	3.35 × 10^{-3}	1.26 × 10^{-4}	1.25 × 10^{-4}	1.19
	3.80 × 10^{-3}	1.41 × 10^{-4}	1.40 × 10^{-4}	0.85
	4.25 × 10^{-3}	1.56 × 10^{-4}	1.56 × 10^{-4}	0.51
	4.86 × 10^{-3}	1.77 × 10^{-4}	1.76 × 10^{-4}	0.23
	5.46 × 10^{-3}	1.97 × 10^{-4}	1.97 × 10^{-4}	0.05
463.15	5.46 × 10^{-3}	9.96 × 10^{-5}	1.01 × 10^{-4}	1.10
463.15	3.50 × 10^{-3}	6.80 × 10^{-5}	6.59 × 10^{-5}	3.09
483.15	5.46 × 10^{-3}	1.28 × 10^{-4}	1.29 × 10^{-4}	0.86
503.15	2.14 × 10^{-3}	6.88 × 10^{-5}	6.75 × 10^{-5}	1.89
503.15	3.95 × 10^{-3}	1.16 × 10^{-4}	1.15 × 10^{-4}	1.03
533.15	2.74 × 10^{-3}	1.06 × 10^{-4}	1.04 × 10^{-4}	1.98
553.15	3.35 × 10^{-3}	1.44 × 10^{-4}	1.42 × 10^{-4}	0.90
553.15	5.46 × 10^{-3}	2.25 × 10^{-4}	2.24 × 10^{-4}	0.67
573.15	3.95 × 10^{-3}	1.88 × 10^{-4}	1.87 × 10^{-4}	0.42
598.15	5.46 × 10^{-3}	2.88 × 10^{-4}	2.86 × 10^{-4}	0.69
603.15	5.46 × 10^{-3}	2.92 × 10^{-4}	2.94 × 10^{-4}	0.62

[a]TDM is the molecular distillation temperature, m_0 is the feed mass flow rate, D_{zexp} is the experimental overall distillate flow rate

obtained from molecular distillation, D_{zcal} is the overall distillate flow rate predicted by using the DESTMOL-P software, AAD % is the Average Absolute Deviation between the experimental and the predicted overall distillation flow rate.

Fig. 13, Fig. 14 and Fig. 15 show the profiles of calculated and experimental D_z according to the molecular distillation conditions described in Table 5. That is, at constant m_0 and variable TDM; at constant TDM and variable m_0 and for variations of both m_0 and TDM, respectively.

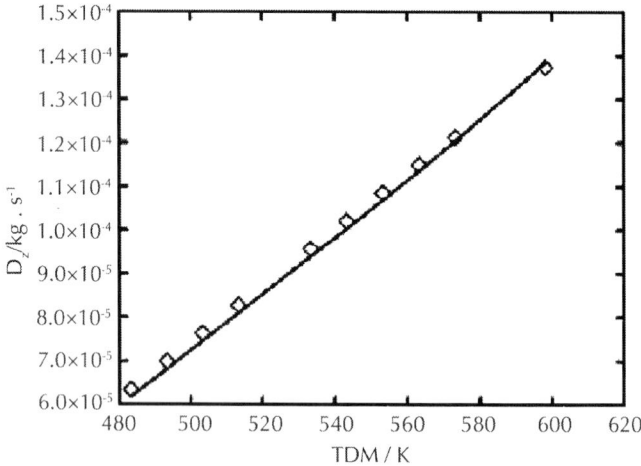

Figure 13: Overall distillate flow rate D_z (kg s^{-1}) profile as a function of molecular distillation temperature for a constant feed mass flow rate. m_0 = 2.4 × 10^{-3} kg s^{-1}; — D_{zcal}; ◊ D_{zexp}.

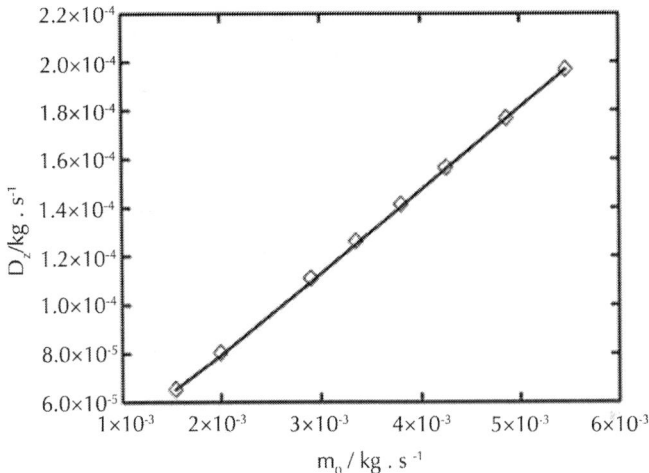

Figure 14: Overall distillate flow rate D_z (kg s^{-1}) profile as a function of feed mass flow rate for a constant molecular distillation temperature. TDM = 598.15 K; — D_{zcal}; ◊ D_{zexp}.

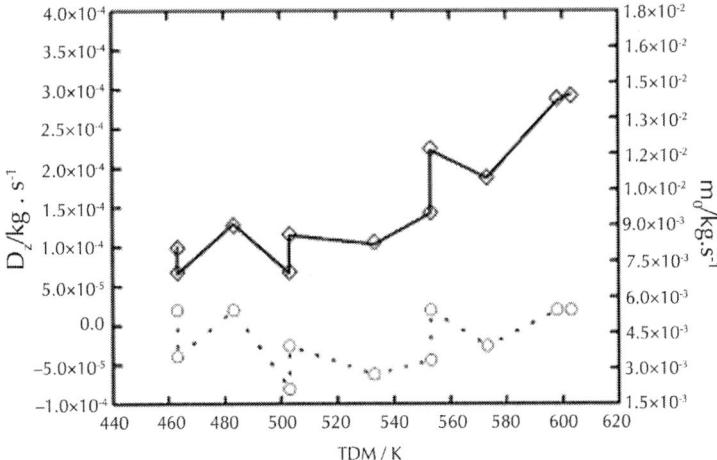

Figure 15: Overall distillate flow rate D_z (kg s^{-1}) profile as a function of molecular distillation temperature and feed mass flow rate. — D_{zcal}; ◊ D_{zexp}, --o-- m_0.

As is observed the prediction made by using the DESTMOL-P is quite approximated of experimental data, with an AAD ranging from 0.02% to 3.09%. Also, it can be perceived that the distribution of AAD values is completely random, and is not observed the propagation errors in the model, i.e. the AAD values are independent of TDM and m_0 values.

CONCLUSIONS

In this work, the modeling and simulation of molecular distillation process of an atmospheric residue were developed. The mathematical model of Kawala and Stephan (1989), the modifications made by Batistella and Wolf Maciel (1996) and the pressure system effect introduced by Sales-Cruz and Gani (2006) were considered to formulated the particular model for the petroleum residue. Values of specific parameters of the process, such as thermal conductivity and pressure system effect were defined for the case study and the DESTMOL-P software was built for the analysis of the process. The results obtained from the simulation of molecular distillation process of a multicomponents mixture split in six pseudocomponents, using a heater evaporator are summarized as follow:

- The temperature of the falling film on the surface, the evaporation rate and the overall distillate flow rate are strongly influenced by the molecular distillation temperature.
- The film undergoes a decreasing in the concentration of more volatile components, mainly from middle to outlet region of the evaporator, where the temperatures are the highest. As a result, a rapid decrease of the film thickness is presented.
- The concentration of not-evaporated pseudocomponents increase in the film along the evaporator.
- All the heat consumed by the evaporated pseudocomponents is continuously supplied from the evaporator wall, thus, a increase of film temperature is presented in the outlet region of evaporator.

The behavior of the variables shown in this work represents the characteristics of the molecular distillation in a heated evaporator, i.e. with high evaporation rates, which reduces the exposure time of the material at operating conditions and the thermal decomposition.

ACKNOWLEDGMENTS

We are grateful to CAPES, FAPESP, CNPq and FINEP for supporting this research. Also, we would like to thank to the CENPES/Petrobras for the performance of HTDS, SARA and physicochemical analysis, involving the Gamma + Sigma crude oil. Also, the Professor Watson Loth of the Chemistry Institute-UNICAMP, for allowing the use of Knauer Vapor Pressure.

REFERENCES

1. Amin, M.B., Beg, S.A., 1993. Kinematic viscosity–temperature behavior of heavy TBP-fractions (455 °C+) of Arabian crude oils. Fuel Sci. Technol. Int. 11, 1425–1439.
2. ASTM D 70, 2008. Annual Book of ASTM Standards, American Society for Testing and Materials. Standard Test Method for Density of Semi-Solid Bituminous Materials (Pycnometer Method). ASTM International, PA. In press.
3. ASTM E 1269, 2005. Annual Book of ASTM Standards, American Society for Testing and Materials. Standard Test Method for Determining Specific Heat Capacity by Differential Scanning Calorimetry. ASTM International, PA. In press.
4. ASTM D 2503, 2008. Annual Book of ASTM Standards, American Society for Testing and Materials. Standard Test Method for Relative Molecular Mass (Molecular Weight) of Hydrocarbons by Thermoelectric Measurement of Vapor Pressure. ASTM International, PA. In press.
5. ASTM D 2892, 2005. Annual Book of ASTM Standards, American Society for Testing and Materials. Standard Test

Method for Distillation of Crude Petroleum (15-Theoretical Plate Column). ASTM International, PA. In press.
6. ASTM D 4124, 2009. Annual Book of ASTM Standards, American Society for Testing and Materials. Standard Test Method for Separation of Asphalt into Four Fractions. ASTM International, PA. In press.
7. Ballesteros, H.J.A., Zuniga, ˜ L.L., Jardini, A., Wolf Maciel, M.R., Maciel Filho, R., Medina, L.C., 2009. Determination of a physical-chemical parameter for modeling process of molecular distillation. Chem. Eng. Trans. 17, 1789–1794.
8. Barros, N.B., Scarminio, I.S., Bruns, R.E., 2003. Como fazer experimentos: pesquisa e desenvolvimento na ciência e na indústria, first ed. UNICAMP, Campinas.
9. Batistella, C.B., Wolf Maciel, M.R., 1996. Modeling, simulation and analysis of molecular distillators: centrifugal and falling film. Comput. Chem. Eng. 20, 19–24.
10. Batistella, C.B., Wolf Maciel, M.R., 1998. Recovery of carotenoids from palm oil by molecular distillation. Comput. Chem. Eng. 22, 53–60.
11. Batistella, C.B., 1999. Tecnologia da destilac¸ão molecular: da modelagem matemática à obtenc¸ão de dados experimentais aplicada a produtos de química fina. Thesis, State University of Campinas.
12. Batistella, C.B., Maciel, M.R.W., Maciel Filho, R., 2000. Rigorous modeling and simulation of molecular distillators: development of a Simulator under condition of non ideality of the vapor phase. Comput. Chem. Eng. 24, 1309–1315.
13. Batistella, C.B., Moraes, E.B., Maciel, Filho, R., Wolf Maciel, M.R., 2002. Molecular distillation, rigorous modeling and simulation for recovering vitamin E from vegetal oils. Appl. Biochem. Biotechnol. 98–100, 1187–1206.
14. Box, G.E.P., Lucas, H.L., 1959. Design of experiments in nonlinear situations. Bioka 46, 77–90.

15. Box, M.J., 1971. An experimental design criterion for precise estimation of a subset of the parameters in a nonlinear model. Bioka 58, 149–153.
16. Brandão Pinto, U., 2002. Use of high temperature simulated distillation analysis for extrapolating true boiling point curve of crude oils. Bol. Técnico Petrobras 45, 343–349.
17. Buzzi-Ferraris, G., Manenti, F., 2009. Kinetic models analysis. Chem. Eng. Sci. 64, 1061–1074.
18. Carnahan, B., Luther, H.A., Wilkes, J.O., 1969. Applied Numerical Methods. John Wiley & Sons, Inc., New York.
19. Chen, F., Cai, T., Zhao, G., Liao, X., Guo, L., Hu, X., 2005. Optimizing conditions for the purification of crude octacosanol extract form rice bran wax by molecular distillation analyzed using response surface methodology. J. Food Eng. 70, 47–53.
20. Cvengros, ̌J., Lutisan, ̌J., Micov, M., 2000. Feed temperature influence on the efficiency of a molecular evaporator. Chem. Eng. J. 78, 61–67.
21. Cvengros, ̌J., Pollák, S., ̌Micov, M., Lutisan, ̌J., 2001. Film wiping in the molecular evaporator. Chem. Eng. J. 81, 9–14.
22. De Andrande Ferreira, A., Radler De Aquino Neto, F., 2005. Simulated distillation in the petroleum industry. Qumica Nova 28, 478–482.
23. Franceschini, G., Macchietto, S., 2008. Novel anticorrelation criteria for model-based experiment design: theory and formulations. AIChE J 54, 1009–1024.
24. Hickman, K.C.D., 1943. High-vacuum short-path distillation—a review. Chem. Rev. 34, 51–106.
25. Hosten, L.H., 1974. A sequential experimental design procedure for precise parameter estimation based upon the shape of the joint confidence region. Chem. Eng. Sci. 29, 2247–2252.
26. Kawala, Z., Stephan, K., 1989. Evaporation rate and separation factor of molecular distillation in a falling film apparatus. Chem. Eng. Technol. 12, 406–413.

27. Lutisan, ˇJ., Cvengros,ˇ J., 1995. Mean free path of molecules on molecular distillation. Chem. Eng. J. 56, 39–50.
28. Lutisan, ˇJ., Cvengros,ˇ J., Micov, M., 2002. Heat and mass transfer in the evaporating film of a molecular evaporator. Chem. Eng. J. 85, 225–234.
29. Maciel Filho, R., Batistella, C.B., Sbaite, P., Winter, A., Vasconcelos, C.J.G., Wolf Maciel, M.R., Gomes, A., Medina, L., Kunert, R., 2006. Evaluation of atmospheric and vacuum residues using molecular distillation and optimization. Pet. Sci. Technol. 24, 275–283.
30. Merdrignac, I., Espinat, D., 2007. Physicochemical characterization of petroleum fractions: the state of the art. Oil Gas Sci. Technol. 62, 7–32.
31. Micov, M., Lutisan, ˇJ., Cvengros,ˇ J., 1997. Balance equations for molecular distillation. Sep. Sci. Technol. 32, 3051–3066.
32. Niederberger, J., Zech, I.A., Da Silva, J.A., Mizutani, F.T., De Sosa, J.S., 2005. 2nd Mercosur Congress on Chemical Engineering, 4th Mercosur Congress on Process Systems Engineering, Rio de Janeiro. PETROX-Petrobras' process simulator.
33. O'Donnell, R.J., 1980. Predict thermal expansion of petroleum. Hydrocarb. Process. 59, 229–231.
34. Pedersen, K.S., Thomassen, P., Fredenslund, A., 1984. Thermodynamics of petroleum mixtures containing heavy hydrocarbons. 1. Phase envelope calculations by use of soave-redlich-kwong equations of state. Ind. Eng. Chem. Process. Des. Dev. 23, 163–170.
35. Perry, R.H., Green, D.W., 1999. Perry's Chemical Engineers' Handbook, seventh ed. McGraw-Hill Companies Inc., New York.
36. Poling, B.E., Prausnitz, J.M., O'Connell, J.P., 2004. The Properties of Gases and Liquids, fifth ed. McGraw-Hill, New York.

37. Sabadini, E., Assano, E.M., Atvars, T.D.Z., 1997. Molecular weight of polyethylene glycols by vapor pressure osmometry: an alternative data treatment. J. Appl. Polym. Sci. 65, 595–600.
38. Sales-Cruz, M., Gani, R., 2006. Computer-aided modelling of short-path evaporation for chemical product purification, analysis and design. Chem. Eng. Res. Des. 84, 583–594.
39. Sbaite, P., Batistella, C.B., Winter, A., Vasconcelos, C.J.G., Wolf Maciel, M.R., Maciel Filho, R., Gomes, A., Medina, L., Kunert, R., 2006. True boiling point extended curve of vacuum residue through molecular distillation. Pet. Sci. Technol. 24, 265–274.
40. Smith, K., 1918. On the standard deviations of adjusted and interpolated values of an observed polynomial function and its constants and the guidance they give towards a proper choice of the distribution of observations. Bioka 12, 1–85.
41. Xubin, Z., Chunjian, X., Ming, Z., 2005. Modeling of falling film molecular distillator. Sep. Sci. Technol. 40, 1371–1386.
42. Zuniga, ˜L.L., 2009. Modelagem e simulacção do processo de destilacção molecular e determinacção experimental aplicado a resíduos pesados de petróleos. Thesis, State University of Campinas.
43. Zuniga, ˜L.L., Lima, N.M.N., Batistella, C.B., Wolf Maciel, M.R., Maciel Filho, R., Medina, L.C., 2008. Correlacção para predizer as massas moleculares de cortes pesados de petróleo. Rev. Petro Química 304, 67–72.
44. Zuniga, ˜L.L., Lima, N.M.N., Wolf Maciel, M.R., Maciel Filho, R., Medina, L.C., Manenti, F., 2010a. Análise reológica de resíduos pesados de petróleo. Aplicacção ao processo de destilacção molecular. Rev. Petro Química 327, 46–53.
45. Zuniga, ˜L.L., Lima, N.M.N., Wolf Maciel, M.R., Maciel Filho, R., Medina, L.C., Embirucu, M., 2011. Correlation for predicting the molecular weight of Brazilian petroleum residues and cuts: An application for the simulation of a molecular distillation process. J. Pet. Sci. Eng. 78, 78–85.

46. Zuniga, ~L.L., Savioli, M.L., Lima, N.M.N., Wolf Maciel, M.R., Maciel Filho, R., Embiruçu, M., Medina, L.C., 2010b. Molecular distillation of petroleum residues and physical–chemical characterization of distillate cuts obtained in the process. J. Chem. Eng. Data 55, 3068–3076.

Chapter 5

Removal of Petroleum Sulfonate from Aqueous Solution by Hydroxide Precipitates Generated from Leaching Solution of White Mud

Lu Cheng[a], Lanlan Ye[a], Dejun Sun[b], Tao Wu[b], and Yujiang Li[a]

[a]Shandong Provincial Key Laboratory of Water Pollution Control and Resource Reuse, School of Environmental Science & Engineering, Shandong University, Jinan 250100, PR China

[b]Key Laboratory of Colloid & Interface Science of Education Ministry, Shandong University, Jinan 250100, PR China

ABSTRACT

Freshly generated hydroxide precipitates (FGHPs), which was prepared by adding the leaching solution of white mud (LSWM) to highly alkaline solution, was used to remove petroleum sulfonate (PS) from aqueous solution. The chemical composition of the white mud and petroleum sulfonate were determined by X-ray fluorescence spectrometry and gas chromatography–mass spectrometry. The surface properties of the FGHPs and PS-FGHPs were characterized by X-ray diffraction, transmission electron microscopy, Fourier transform infrared spectroscopy and zeta potential analyzer. The FGHPs displayed excellent treatment efficiency for PS at pH 12.0. The maximum equilibrium removal efficiency of PS was reached within 60 s. The maximum adsorption capacity of FGHPs for PS was 3798.06 mg/g at 303 K and pH 12.0. The Langmuir isotherm was the best choice to describe the adsorption behavior. The kinetic data fitted the pseudo-second-order kinetic model. Thermodynamic parameters suggested that the adsorption of PS onto FGHPs occurred via physisorption and was exothermic. Electrostatic attraction and hydrogen bonding were the main adsorption mechanisms. Moreover, adhesion and cohesion also strong affected the co-precipitation/adsorption process. Liquid bridges via hydrogen bonding, adhesive and cohesive forces linked up with $MOH^+/M(OH)_2$ particles and the surfactant molecules to form a three-dimensional network structure and lead to deposition.

INTRODUCTION

Surfactant wastewater is one of the major pollution sources of a receiving water body around the world [1]. In the surfactant wastewater produced by the households, surface active agents or surfactants invariably exist in significant amounts due to detergents used for all kinds of washings [1]. Surfactants have also been widely used in textiles, fibers, food, paints, cosmetics, pharmaceuticals, mining, oil recovery and pulp and paper industries [1] and [2]. The synthetic surfactants are of three major types: anionic, nonionic, and cationic. Anionic surfactants are the major class of surfactants and have been used widely in detergent formulations[3] and [4]. Petroleum sulfonate (PS) surfactants differ from relatively homogeneous conventional surfactants in that they are mixtures of sulfonated alkyl–aryl petroleum products and free mineral oils [5]. Polymers and PS surfactants are used as effective "pusher fluids" to enhance oil recovery and the PS surfactants are therefore used to improve the flooding efficiency of crude oil and widely used in exploitation [6]. Some surfactants are toxic; others are not, depending on dose, chemistry, receptors, etc. They may cause foaming in rivers and reduce the quality of water [3]. Surfactants cause short and long-term changes in ecosystems[3]. PS surfactants removal from water environment before they are contacted with natural water bodies is very important.

Many conventional methods exist for the removal of anionic surfactants from water environment. These involve chemical and electrochemical [7] and [8], membrane separation [9], photocatalytic degradation [10], adsorption [11] and biological treatment processes [12]. However, anionic surfactants having high degree of hydrophilic nature and poor biological activity, biological degradation of anionic surfactants occurs too slowly. These methods have been found to be limited, and none of them were successful in completely removing anionic surfactants from aqueous solution [5], [7], [8], [9], [10], [11], [12] and [13]. Hence it would be of much practical and academic interest to investigate surfactant wastewater treatment using new methods with low cost and high efficiency.

Co-precipitation/adsorption processes have high adsorption abilities and short adsorption equilibrium times and may be suitable for surfactant wastewater treatment [14], [15], [16] and [17]. For example, when metallic ions precipitate in alkaline conditions, the freshly formed hydroxides are fine particles with a larger specific surface area and higher surface free energy. As the hydroxide particles are formed, they incorporate and adsorb pollutants, so the adsorption ability can be enhanced significantly by co-precipitation/adsorption processes [14], [15], [16], [17] and [18].

Industrial wastes are remarkable adsorbents in wastewater treatment. They provide several advantages such as their low-cost, abundant availability and possibility for sludge disposal [19], [20] and [21]. The major industrial process for the production of soda ash is the ammonia–soda process. In this process, a large amount of fine-sized solid wastes known as white mud is produced. Since white mud is characterized by high water content, high alkalinity and high concentrations of soluble Ca^{2+}, Mg^{2+} and Cl^- ions [5] and [22], it is difficult to deal with or reuse. Common treatment methods for white mud include landfill disposal or discharge to oceans [23], which causes a potential threat to groundwater, soil and the marine environment. Therefore, a new method is needed to reuse these problematic materials [22], [23] and [24].

In this work, we attempt to reduce raw materials and operating costs for PS removal by industrial wastes treating PS-containing wastewater. Co-precipitation/adsorption process has been used to remove PS from aqueous solution. When the leaching solution of white mud (LSWM) is added to alkaline solution, metal hydroxides are precipitated. This freshly generated hydroxide precipitates (FGHPs) has positive surface charge, which can be used as adsorbent to attract negatively charged PS surfactant. The effect of pH, initial PS concentration, adsorbent dosage, contact time and temperature were investigated. Adsorption isotherms and kinetic data were obtained and thermodynamic parameters were determined. The adsorption mechanism between FGHPs and PS was also discussed.

MATERIALS AND METHODS

Materials

Alkaline white mud was obtained from the Shandong Aluminum Plant, dried at 103–105 °C overnight, disaggregated and sieved to

200 mesh prior to use. Its chemical composition as determined by X-ray fluorescence spectrometry (XRF, ZSX Primus II, Rigaku Corporation, Tokyo, Japan) is shown in Table 1. The white mud was dispersed in 1.2 mol/L hydrochloric acid with a solid content of 0.1268 g/mL for 12 h and filtered to obtain the leaching solution.

Table 1: Chemical constituents of white mud

Constituent	$CaCO_3$	$CaSO_4$	$Mg(OH)_2$	Fe_2O_3	Al_2O_3	SiO_2	CaO	$CaCl_2$	$NaCl$
Percentage by weight (%)	36.47	6.10	24.30	1.84	2.99	9.69	4.17	7.86	2.38

The PS surfactant used in this present work was obtained from the ShengLi oilfield in China. The petroleum sulfonate surfactants are mixtures of petroleum sulfonates and free mineral oils that form stable compound. The active components of PS sample were refined by solvent-extraction method to extract PS surfactant. Pentane, absolute alcohol, isopropyl alcohol and petroleum ether (boiling range 308–338 K) could be used as extracting solvent. In this work, we used isopropanol–water mixtures (1:1, v/v) as extracting solvent to purify PS sample for further experiments. The active components of PS sample are characterized by gas chromatography–mass spectrometry (GC–MS). GC–MS analysis were performed with a Thermo Finnigan TRACE (Thermo Finnigan, San Jose, CA, USA) coupled with a TRACE MS plus (EI 70 eV) from the same manufacturer. The analyses were carried out using different fused silica capillary columns (30 m × 0.25 mm i.d.; film thickness 0.25 μm) of different polarities (DB-5 and HP-Innowax) from Agilent Company (Palo Alto, CA, USA). The oven temperature was programmed from 50 to 250 °C at 3 °C/min and held isothermal for 10 min. Injector and interface temperatures were 220 and 250 °C, respectively. The carrier gas was helium at a flow rate of 1 mL/min. Diluted samples (1/10 in ether) of 1.0 μL were injected manually and the split ratio was adjusted to 40:1. The components were identified by comparing their mass spectra with those in the NIST98 GC–MS library and these in the literature (Huang et al.

[2]), as well as by comparing their retention times. The components were 53.97% PS, 39.69% free mineral oil and 4.13% inorganic salts, with a recovery rate of 96.8% [5]. NaOH/HCl was used to adjust the solution pH.

All other chemical reagents were of analytical grade and were used without further purification.

Adsorption Experiments

Adsorption experiments for PS were carried out using a batch equilibrium technique. A 5000 mg/L stock solution was prepared in deionized water, followed by successive dilution to obtain the necessary concentrations. The solution pH was adjusted to the required values using 0.5 mol/L HCl and 2.0 mol/L NaOH solution.

Adsorption of PS was performed by adding a given amount of leaching solution into 250-mL beakers containing 100 mL PS solution at the required concentration and pH. The mixture was stirred continuously at 350 rpm using a Jintan JJ-1 motor stirrer for a given contact time for adsorption. After stirring, the mixture was allowed to settle under static conditions for 10–15 min. The mixture was centrifuged at 8000 rpm (LG10-2.4A, Beijing Medical Centrifuge Factory, China) for 15 min to separate the unsettled small particles from the supernatant after settling. The concentrations of PS in the supernatant were determined using a UV–visible spectrophotometer (UV-1601, Shimadzu, Japan) at wavelength of 652 nm corresponding to the maximum adsorbance. The effect of pH on PS removal was examined at 303 K, an initial PS concentration of 200 mg/L, and pH values were varied; the effect of adsorbent dosage at 303 K and pH 12.0 was studied by varying the adsorbent dosage (0.50–2.50 g/L) at initial PS concentration in the range of 100–500 mg/L. The effect of temperature on adsorption was assessed by varying the temperature from 303 K to 333 K. Adsorption isotherm studies were then conducted by contacting a fixed amount (1.50 g) of FGHPs with the PS concentration of 200 mg/L. Two kinetic models were used to fit the experimental data at different adsorbent dosage (0.5–2.5 g/L) and fixed PS concentration at 303 K. The removal efficiency of PS, (%), which can be calculated from the concentration of PS before (C_0) and after adsorption (C_e):

$$\eta = (C_0 - C_e)/C_0 \times 100\% \tag{1}$$

The adsorption capacity, q_t, is determined from following equation:

$$q_t = (C_0 - C_e)V/W \tag{2}$$

where W (g) is the mass of adsorbent, and V (L) is the volume of solution.

Characterization of Adsorbent

The LSWM was added into a NaOH solution (pH = 12.0) to produce hydroxide precipitates which was used as the adsorbent to adsorb the PS surfactant. Transmission electron microscopy (TEM, TEM-100X II, JEOL, Tokyo, Japan) was employed to analyze the surface morphology of the precipitates. The average hydrodynamic particle diameter of the freshly generated precipitates was measured by dynamic light scattering (DLS, BI-200SM/BI-9000, Brookhaven Instrument Co., New York, USA). The isoelectric point (IEP) of hydroxide precipitates was determined by a laser electrophoresis zeta potential analyzer (Zetasizer III from Malvern instruments). The analyses were carried out in aqueous solution at ambient temperature, and the zeta potential was an average of measurements. PS surfactant and the freshly generated precipitates before and after adsorption were characterized by Fourier transform infrared spectroscopy (FT-IR, Bruker Tensor 27 spectrometer, USA) and X-ray diffraction (XRD, D/max 2550V, Rigaku, Japan) with mono-chromated Cu K radiation.

RESULTS AND DISCUSSION

Characterization of Leaching Solution and Freshly Generated Hydroxides

The main constituents of the white mud as measured by XRF are given in Table 1. White mud consists primarily of $CaCO_3$, $CaSO_4$, $CaCl_2$, $Ca(OH)_2$, dead-burnt CaO, $Mg(OH)_2$ and SiO_2. We used X-ray diffraction to investigate the components and crystal structures of the white mud, LSWM, FGHPs and PS-FGHPs. XRD is the basic technique to determine the bulk structure and composition of materials with crystalline structure. It can be seen that there were numerous sharp peaks in the XRD pattern of the white mud. The shape of the sharp diffraction peaks also indicates that the white mud is well crystallized. As shown in Fig. 1a, white mud is mainly composed of amorphous phase and crystal phase present in the form of calcium carbonate and magnesium hydroxide (JCPDS cards: No. 47-1743 and No. 74-2220). Moreover, crystalline content is also present in white mud in the form of calcium chloride and calcium sulfate. As shown in Fig. 1b, LSWM is mainly present in the form of calcium chloride and magnesium chloride (JCPDS cards: No. 25-0515 and No. 26-1053). Such difference in crystal phases may imply the distinctive adsorption characteristics of adsorbents. The XRD pattern of the precipitates is shown in Fig. 1c. All diffraction peaks corresponding to (0 0 1), (1 0 1) and (1 1 0) reflection of $Mg(OH)_2$ were observed for the precipitates, which are well consistent with the literature data (JCPDS card: No. 74-2220). This may suggest that the precipitates had the basic structures of $Mg(OH)_2$. Furthermore, as shown in Fig. 1d, the XRD peaks of PS-FGHPs are reduced after the adsorption of PS onto FGHPs comparing with neat FGHPs, since the FGHPs could hinder the mobility of PS molecules through the hydrogen bonding, adhesion of PS molecules onto FGHPs particles will impose, to a certain extent, a possible confinement effect on PS molecules diffusion, leading to a great increase in crystal growth. On the other hand, electrostatic attraction between positive FGHPs particles and negative PS molecules, resulting in a great increase in overall crystallization rate, it is reasonable to believe that the crystal growth will be enhanced, which finally formed the poorly crystalline phase. The intensity of the peaks, particularly (1 0 1) and (1 1 0), decreased in comparison with the neat FGHPs. Therefore, the main components of the hydroxide precipitates, which are generated in alkaline conditions, are: $Ca(OH)_2$, $Mg(OH)_2$, $Fe(OH)_3$ and $Al(OH)_3$. The initial hydraulic diameter of the freshly generated precipitates as measured by DLS was 6.0 nm on average. The particles therefore have a relatively high specific surface area that is available for the adsorption of PS surfactant. Many hydroxide precipitates are observed to aggregate into flocs (Fig. 2), and precipitates are connected with each other through adhesion and cohesion. The loosely-packed hydroxide precipitates extend into the aqueous phase and serve as bridges between neighboring precipitates; the bridges form a three-dimensional network structure and aid adsorption.

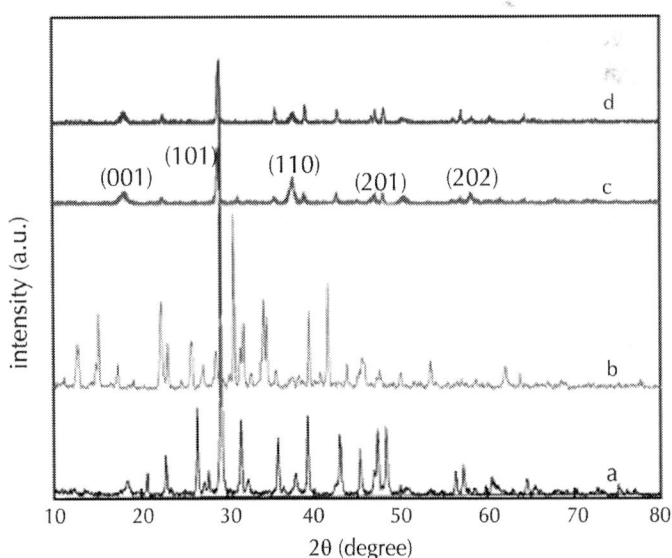

Figure 1: XRD patterns of (a) white mud, (b) LSWM, (c) FGHPs and (d) PS-FGHPs.

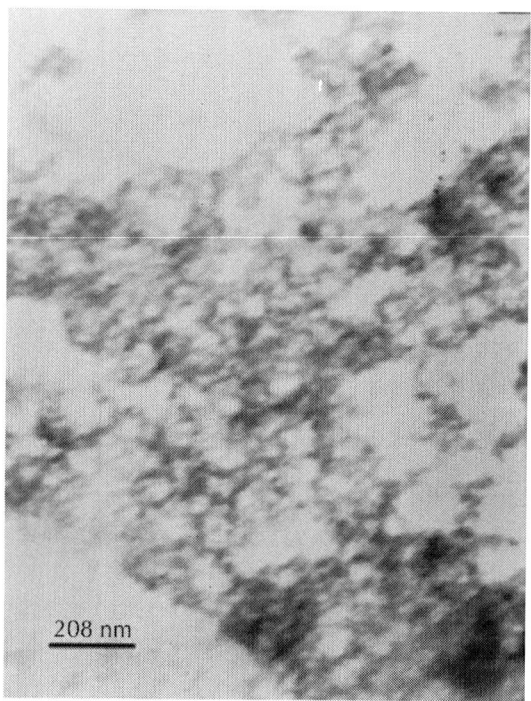

Figure 2: Transmission electron microscopy micrograph of FGHPs.

On the other hand, since various metallic ions are simultaneously present in the system, the diffusion rate of various metallic ions were different, the time needed for the growing units to add to the crystal lattice is different, which lead to the formation of an amorphous crystalline structure with higher surface free energy. In this case, the adsorption of PS onto FGHPs is enhanced.

FT-IR spectroscopy is a sensitive tool for examining the molecular environment of the surfactant (PS) within the PS-FGHPs. The infrared spectra of the FGHPs with and without adsorbed PS are shown in Fig. 3. The spectrum of the pure FGHPs display bands at 3696, 3441, 1447, 874 and 441 cm^{-1}. A sharp and intense peak at 3696 cm^{-1} is assigned to a free hydroxyl group [25], while the broad band at 3441 cm^{-1} may be ascribed to O-H stretching vibration of hydroxyl groups and water molecules in the interlayer including H-bonded and physically adsorbed water [26]. The bands in the low wavenumber region (<900 cm^{-1}) of the spectrum can be interpreted as lattice vibrations of FGHPs, such as M-O-H (874 cm^{-1}) and O-M-O (441 cm^{-1}) vibrations are observed [27]. The adsorption band at 1447 cm^{-1} shows the presence of carbonate as a unidentate complex with aluminum [28]. The spectrum of the PS surfactant display bands at 2922, 2855, 1464, 1188, 1047, 828 and 583 cm^{-1}. The bands at ~2930 and ~2850 cm^{-1} are attributed to CH_2 asymmetric stretching vibration and CH_2 symmetric stretching vibration, respectively. They are sensitive to change in the gauche conformer ratio and chain–chain interactions [28] and [29]. For the PS surfactant, CH_2 asymmetric and symmetric stretching vibrations were found at 2922 and 2855 cm^{-1} (Fig. 3). The band at 1464 cm^{-1} for the surfactant molecules is attributed to C-H in-plane bending vibrations. The $-SO_3^{2-}$ asymmetric and symmetric stretching vibrations were found at 1188 and 1047 cm^{-1}. The band at 828 cm^{-1} and 583 cm^{-1} are attributed to S-O-S stretching vibration and C-C in-plane bending vibration [28], [29] and [30]. After the adsorption of PS, CH_2 asymmetric and symmetric stretching vibrations were found at 2924 and 2854 cm^{-1}. The band at 2513 cm^{-1} is attributed to CO_2, which is from the decomposition of labile functional groups[28], [29], [30] and [31]. When PS is adsorbed on the FGHPs, shifts in the position of the bands are observed. The band at 3696 shift to 3698 cm^{-1}; 3441 shift to 3415 cm^{-1}; 2922 shift to 2924 cm^{-1}; 2855 shift to 2854 cm^{-1}; 1464 shift to 1463 cm^{-1}; 1188 shift to 1180 cm^{-1}; 1047 shift to 1040 cm^{-1}; 874 shift to 873 cm^{-1}, respectively. Shifts in these bands indicate that the hydrophilic sulfonic groups of PS molecules reacted with FGHPs particles. FGHPs particles provide ideal sites for PS to build up by hydrogen bonding and enhance the adsorption of PS molecules onto FGHPs [29].

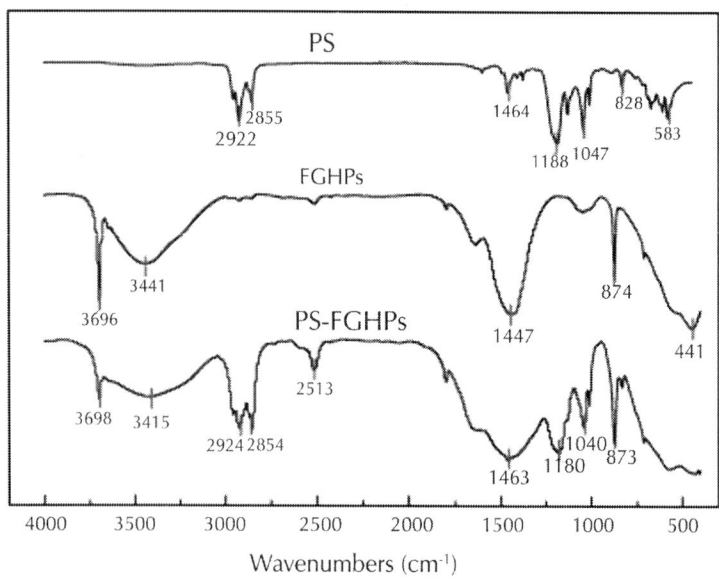

Figure 3: FT-IR spectra of PS, neat FGHPs and after the adsorption of PS onto FGHPs.

Effect of pH

The initial pH of the PS solution is an important parameter that controls the adsorption process and particularly the adsorption capacity. The effect of pH on PS removal efficiency is shown in Fig. 4. The initial pH ranges from 10.5 to 13.0. The removal efficiency increased with pH until the pH reached 12.0, and then the removal efficiency of PS decreased with further increasing pH. The main solubility product constants of the precipitates used for adsorption are:

$Ca(OH)_2 \quad pK_{sp} = 5.26$
$Mg(OH)_2 \quad pK_{sp} = 10.74$
$Fe(OH)_3 \quad pK_{sp} = 37.40$
$Al(OH)_3 \quad pK_{sp} = 32.89$

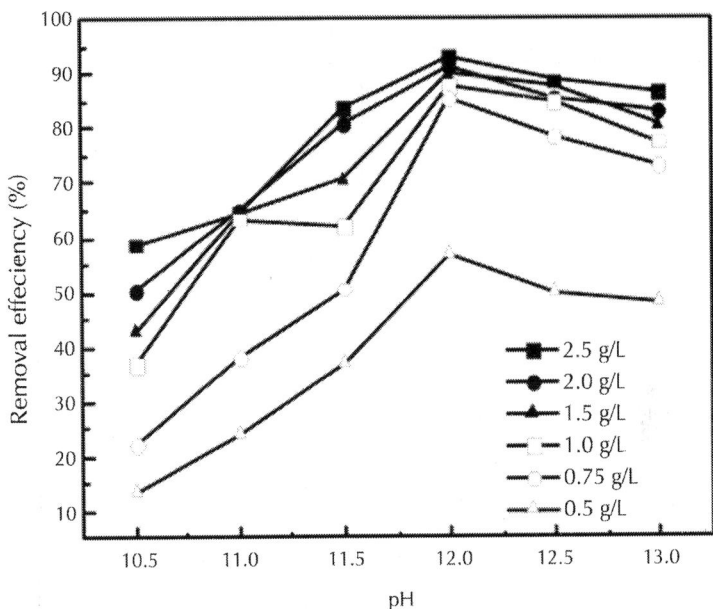

Figure 4: Effect of pH and adsorbent dosage on PS removal efficiencies. Experimental conditions: initial PS concentration = 200 mg/L, T = 303 K and contact time = 60 s.

At the lower initial pH conditions, only $Fe(OH)_3$ and $Al(OH)_3$ can form, so the adsorption ability is limited because of the low content of Fe^{3+} and Al^{3+} in the leaching solution of white mud (Table 1). In contrast, when the pH is high enough, large amounts of $Mg(OH)_2$ and $Ca(OH)_2$ crystals begin to form. Therefore, the removal efficiency increase significantly. An intermediate product, MOH^+ can be formed during the process of generating $M(OH)_2$ crystals[19] and [32], which can be attributed to the increasing OH^- concentration. The IEP of hydroxide precipitates is around 12.2 (Fig. 5), and with increasing pH, the polarity of the FGHPs changes from positive to negative. As the pH increases and exceeds the IEP, the number of positively charged sites decreases and the number of negatively charged sites increases. [24], [32], [33] and [34]. The positive charge develops on the surface of adsorbent and may be written as:

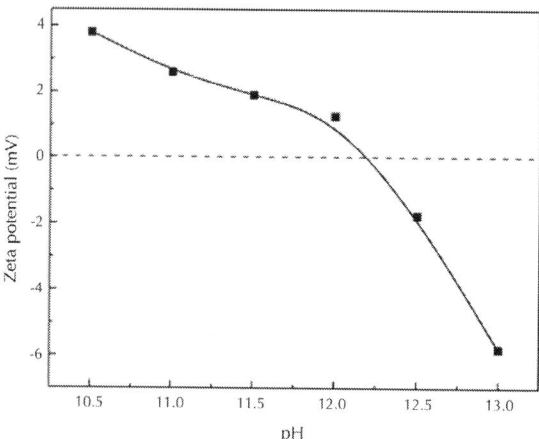

Figure 5: Zeta potential of FGHPs as a function of pH in 0.01 M NaCl solution at 25 °C.

Negatively charged sites on the adsorbent do not favor the adsorption of anionic surfactant due to the electrostatic repulsion. Besides, at high pH, due to the presence of excess hydroxyl ions competing with the anionic surfactant for the adsorption sites. Therefore, at pH >12.0, a repulsive force is expected and a slow reduction in PS adsorption. An initial pH of 12.0 was chosen in subsequent adsorption experiments.

Effect of Adsorbent Dosage

The effect of adsorbent dosage on PS removal efficiency is shown in Fig. 6. The initial PS concentrations ranged from 100 to 500 mg/L. The adsorbent dosage was increased from 0.50 to 2.50 g/L at pH 12.0. The removal efficiency increased rapidly with increasing adsorbent dosage, and then reached a plateau. The PS removal efficiency decreased from 90.79% to 87.96% with the increase of initial PS concentration from 100 to 500 mg/L. When the adsorbent dosage increased to 1.50 g/L, with further increasing the adsorbent dosage did not improve the PS removal efficiency significantly. This phenomenon may be associated with possible growth mechanisms of hydroxide precipitates. The possible growth mechanisms of the hydroxide precipitates is the formation process may be composed of three stages: nucleation, growth and assembly [35]. The first stage is the nucleation process, the initial reaction between the M^{2+} and hydroxyl group, which is too rapid for nuclei generation. The concentration of M^{2+} plays an important role in controlling nucleation and growth. According to the sp^3d^2 hybrid orbital of M^{2+}, which has an empty orbital and is ready to accept an electron pair, M^{2+} ions act as an electron pair acceptor and can be viewed as Lewis acid [36]. While hydroxyl groups have excess electrons act as an electron pair donor and can be viewed as Lewis base. Therefore, strong electrostatic attraction and acid–base neutralized interactions exist between M^{2+} and hydroxyl group. The second stage is MOH^+ nanoplate formation. The hydroxyl group of MOH^+ may interact with sulfonic of PS surfactant are stacked together by hydrogen bonding. On the other hand, a comprehension resulted from relevant cohesive mechanism governing aggregation of micrometer-sized particles, including capillary cohesion for MOH^+ particle–particle, capillary liquid bridge cohesion and capillary adhesion for MOH^+particle–surface of surfactant, leads to the formation of nanoplate through the oriented cohesion and adhesion [37], [38] and [39]. Geometries of capillary cohesion for particle–particle and adhesion for particle–surface of surfactant with a liquid bridge are shown in Fig. 7. The adsorption amount of PS increased rapidly during this process. The third stage is self-assembly process of nanoplates via hydrogen bonding (aggregation), leads to the formation of $M(OH)_2$. All self-assembling systems are driven by some principle of energy minimization [32] and [40]. Electrostatic attraction and hydrogen bonding are strong enough to hold the nanoplates of MOH^+ together [12] and [35]. The co-precipitation/adsorption process occurs during the possible growth of complex $M(OH)_2$ structures. The adsorbent dosage plays an important role in the crystalline morphology.

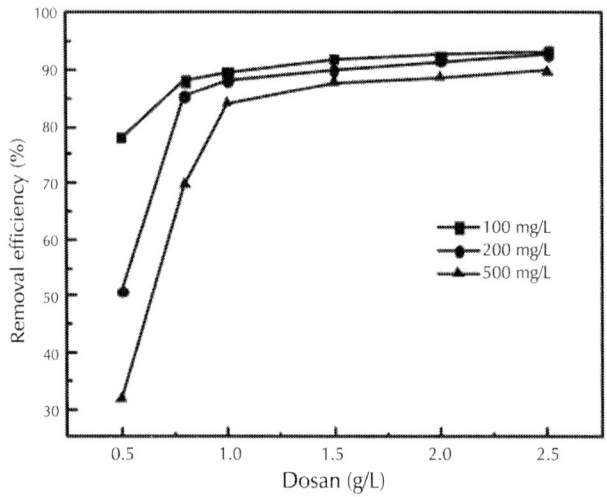

Figure 6: Effect of adsorbent dosage and initial PS concentration on PS removal efficiencies. Experiment conditions: initial pH = 12.0, T = 303 K and contact time = 60 s.

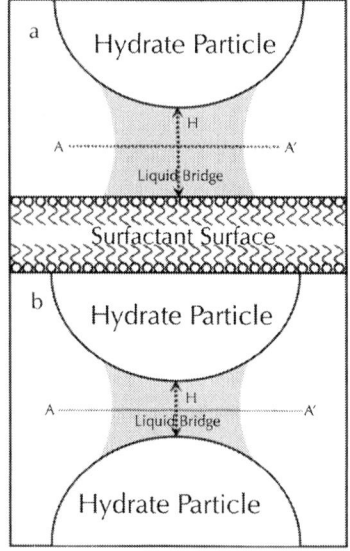

Figure 7: Schematic of capillary (a) adhesion for particle–surface and (b) cohesion for particle–particle geometries with a liquid bridge, in which AA' is the plane of symmetry, H is the liquid bridge height, adapted from

Aman et al. [35] and [36].

Effect of Contact Time and Initial PS Concentration

The relationship between adsorption of PS and contact time was investigated to identify the rate of PS removal. The effect of contact time and initial PS concentration on the removal efficiencies of are shown inFig. 8. The amount of adsorbed PS increased with the increase in contact time and then reached equilibrium at 60 s. This is a rapid process compared with other adsorptions [41], [42] and [43].

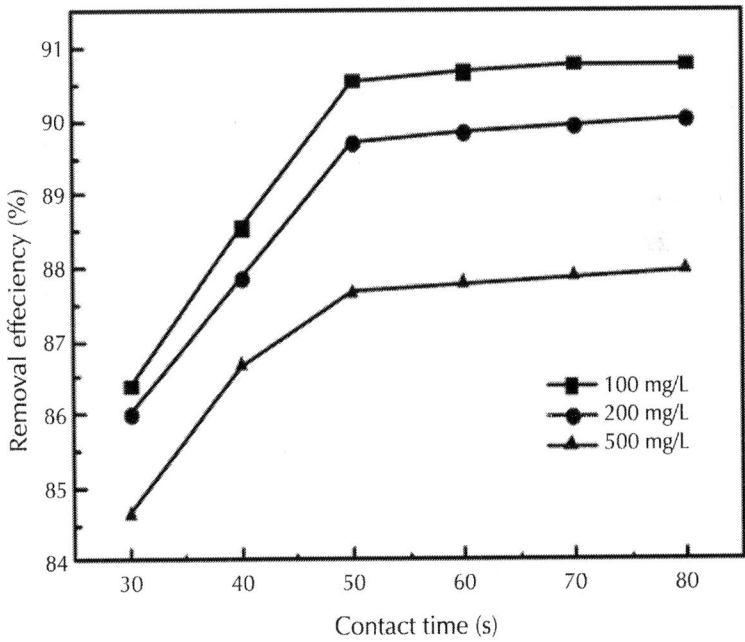

Figure 8: Effect of contact time and initial concentration on PS removal efficiencies. Experimental conditions: initial pH = 12.0, T = 303 K and adsorbent dosage = 1.50 g/L.

Because of the presence of water in bulk powder materials,

and the presence of water vapor between particles and surfaces, capillary condensation can occur because of the pressure difference that exists across the curved interface [44]. Capillary forces have been focused mainly on the condensation of water vapor from the humid atmosphere to form a liquid bridge [44] and [45]. In any atmosphere, capillary condensation between particles will result in an increased attractive interaction [44], [45] and [46]. The capillary forces are greatest when particles are hydrophilic and they are in contact [45], [46] and [47]. Capillary forces will increase the cohesive interaction between particles and enhance agglomeration [47], [48] and [49]. Capillary condensation plays an important role in particles cohesion. The liquid bridges through hydrogen bonding, adhesive and cohesive forces are linked up with the hydrate $MOH^+/M(OH)_2$ particles and surfactant molecules to form a three-dimensional network structure and lead to deposition. On the other hand, adhesive forces between hydrate $MOH^+/M(OH)_2$ particles and surfactant surfaces. The adhesive forces are strongly dependent on the presence of water in the system. The adhesive forces are highest when the surfactant surfaces are water-wet. When a crystal grows from a supersaturation solution, free surfactant molecules diffuse from the bulk phase to the solid–liquid interface region. Surfactant molecules are adsorbed on the $MOH^+/M(OH)_2$ particles via electrostatic attraction and hydrogen bonding. With surfactant molecules to form bonds with the surface groups of the crystal, the surfactant molecules integrate completely into the crystal lattice. As the diffusion rate is higher than that of surface integration, the growing units could not get enough time to add to the crystal lattice. At the same time, various metallic ions are simultaneously present in the system, which leads to an amorphous crystalline structure with higher surface free energy. The adhesive forces increase slightly with increasing surface free energy of the particles [44], [45], [46], [47], [48] and [49]. The larger forces may indicate that hydrate $MOH^+/M(OH)_2$ particles are easily adhered to the surfactant molecules. Both cohesion (particle–particle) and adhesion (particle–surface) play important roles in controlling the nucleation and growth of $M(OH)_2$ particles. Higher surface free energy and larger forces will contribute to the enhanced adsorption of PS surfactant. Moreover, capillary forces draw water from the bulk phase to form liquid bridges between particles and surfaces. Liquid bridges through hydrogen bonding, adhesive and cohesive forces are linked up with the hydrate $MOH^+/M(OH)_2$ particles and surfactant molecules to form a three-dimensional network structure and lead to deposition. Therefore, the adsorption equilibrium time could be much shorter.

The initial concentration provides an important driving force to overcome all mass transfer resistances of the PS between the aqueous phase and solid phase [44]. Experimental results for the adsorption of PS onto FGHPs at various initial concentrations are shown in Fig. 8. The PS removal depends on the concentration of the PS. The percentage of adsorption efficiency decreases with increase in initial PS concentration in the solution. However, the actual amount of PS adsorbed per unit mass of adsorbent increases with increase in PS concentration. Moreover, the initial PS concentration did not affect the equilibration time significantly.

Adsorption Isotherms

Adsorption isotherm analyses reveal how the adsorption molecules distribute between liquid phase and solid phase when the adsorption process reaches an equilibrium state. The analysis of the adsorption isotherm data by fitting the data to different adsorption models is an important step to determine a suitable model that can be used for designing purposes. Two classical adsorption models, i.e., Langmuir and Freundlich, are the most frequently employed models. In this work, two models were used to describe the relationship between the amount of PS adsorbed and its equilibrium concentration in solution at pH 12.0 and different temperatures.

The well-known Langmuir model is expressed by:

$$q_e = \frac{K_L q_m C_e}{1 + K_L C_e} \tag{4}$$

where q_e (mg/g) is the equilibrium PS concentration on the adsorbent, C_e (mg/L) is the equilibrium PS concentration in solution, q_m (mg/g) is the monolayer capacity of the adsorbent, and K_L (L/mg) is the Langmuir adsorption constant which relates to the adsorption energy [50]. The Langmuir equation is applicable to homogeneous adsorption, where the adsorption of each adsorbate molecules onto the surface has equal adsorption activation energy.

The Freundlich equation is given by:

$$q_e = K_F C_e^{1/n} \tag{5}$$

where q_e (mg/g) is the equilibrium PS concentration on the adsorbent, C_e (mg/L) is the equilibrium PS concentration in solution, and K_F and n are the Freundlich constants that are characteristic of the system, and are indicators of adsorption capacity and adsorption intensity, respectively. The Freundlich equation is employed to describe heterogeneous systems and reversible adsorption and is not restricted to the formation of monolayers [50].

Fig. 9 shows the adsorption isotherms of PS onto the adsorbent at different temperatures and the fitted curves using the two equilibrium isotherms. The parameters for the two isotherms obtained from experimental data are presented in Table 2. As seen that the adsorption equilibrium obtained from the Langmuir isotherm is quite close to the experimental data and the regression coefficients are better than those obtained from the Freundlich isotherm. Therefore, the Langmuir isotherm was the best choice to describe the adsorption behavior.

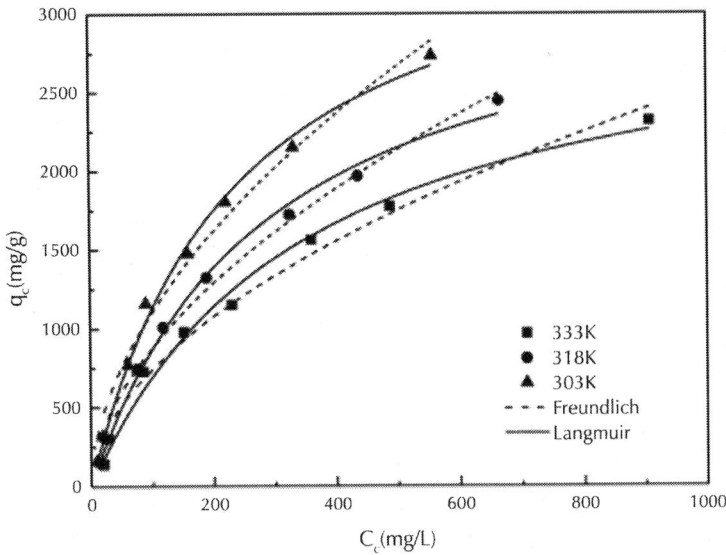

Figure 9: Adsorption isotherms of PS onto FGHPs. Experimental conditions: initial pH = 12.0, T = 303 K and adsorbent dosage = 1.50 g/L.

Table 2: Freundlich and Langmuir isotherms parameters for adsorption of PS onto FGHPs

T (K)	C_0 (mg/L)	Freundlich isotherm			Langmuir isotherm			
		K_F	n	R^2	q_m (mg/g)	K_L (L/mg)	R_L	R^2
303	200	89.05	1.8	0.9831	3798	0.0042	0.5435	0.9945
318	200	69.38	1.8	0.9905	3381	0.0034	0.5952	0.9925
333	200	64.72	1.9	0.9814	3132	0.0029	0.6329	0.9894

For the Langmuir isotherm, a method has been adapted to calculate the dimensionless separation factor (R_L)[51], which determines the favorability and the shape of the isotherm of the adsorption process by applying the equation:

$$R_L = \frac{1}{1 + K_L C_0}$$

(6)

where K_L is the Langmuir constant and C_0 is the initial concentration. The value of R_L indicates whether the isotherm is unfavorable ($R_L > 1$), linear ($R_L = 1$), favorable ($0 < R_L < 1$), or irreversible ($R_L = 0$) [51]. The calculated values of separation factor for the adsorbent at different temperatures are presented in Table 2. All the values of R_L are less than unity, which confirms the favorable adsorption process. The maximum adsorption capacity (q_m) for PS onto FGHPs was obtained using the Langmuir model; q_m was 3798.06 mg/g at 303 K in this work. Compared with other adsorbents [52], [53] and [54], FGHPs displayed superior adsorption capacities. This is because the removal mechanism of PS by FGHPs is a co-precipitation/adsorption process that involves a combination of electrostatic attraction and hydrogen bonding. Moreover, adhesive and cohesive forces also strong affect co-precipitation/adsorption process.

The results in Fig. 9 and Table 2 also indicate that the adsorption capacities decreased significantly with increasing temperature, which indicates that the adsorption process is exothermic in nature, and that a low solution temperature contributes to the adsorption of PS onto FGHPs.

Adsorption Kinetics

The kinetics of adsorption is one of the most important parameters for determining the adsorption mechanism and investigating the efficiency of adsorbent for the removal of pollutants [55]. In this study, two kinetic models, pseudo-first-order and pseudo-second-order models were used to test experimental data and predict the adsorption kinetics.

The pseudo-first-order-kinetic model [55] is given as:

$$\ln(q_e - q_t) = \ln q_e - k_1 t \qquad (7)$$

where q_e (mg/g) and q_t (mg/g) are the amount of PS adsorbed by FGHPs at equilibrium conditions and at time t (min), respectively. k_1 (1/min) is the equilibrium rate constant of pseudo-first-order adsorption [55]. k_1 and q_e are determined from the slope and intercept of the plot of $\ln(q_e - q_t)$ versus t respectively.

The pseudo-second-order kinetic model [56] is expressed as:

$$1/q_t = 1/k_2 q_e^2 + 1/q_e \qquad (8)$$

where q_e (mg/g) and q_t (mg/g) are the same as for the pseudo-first-order parameters, k_2 is the equilibrium rate constant of pseudo-second-order adsorption (g/mg min). Values of k_2 and q_e are obtained from the slope and intercept of the plot of t/q_t versus t, respectively.

A comparison of results with the correlation coefficients is given in Table 3. The calculated $q_{e,cal}$ values obtained from the pseudo-first-order kinetic model do not give reasonable values, which are too low compared with experimental $q_{e,exp}$ values. This finding suggests that the adsorption of PS onto FGHPs is not a pseudo-first-order reaction. On the contrary, the results present an ideal fit to the pseudo-second-order kinetic for adsorbent with the extremely high $R^2 > 0.999$ (Table 3). A good agreement with this adsorption model is confirmed by the similar values of calculated $q_{e,cal}$ and the experimental data $q_{e,exp}$ for all cases. The rate constant, k_2 increased with increasing adsorbent dosage. This implied that the number of active sites increased on the adsorbent surface as parallel to increasing adsorbent dosage. It is clear that the kinetic of PS adsorption onto FGHPs follows the pseudo-second-order model.

Table 3: Adsorption parameters of kinetic for the adsorption of 200 mg/L PS onto FGHPs at 303 K

Dosage (g/L)	$q_{e,exp}$ (mg/g)	Pseudo-first-order			Pseudo-second-order		
		$q_{e1,cal}$ (mg/g)	k_1 (1/s)	R^2	$q_{e2,cal}$ (mg/g)	k_2 (g/mg s)	R^2
0.5	204.18	94.77	0.0861	0.9281	210.53	0.0024	0.9999
1.0	176.36	87.32	0.0866	0.9224	182.48	0.0026	0.9998
1.5	120.32	79.98	0.0867	0.9133	125.94	0.0027	0.9997
2.0	92.03	75.82	0.0875	0.9002	97.46	0.0028	0.9995
2.5	74.46	71.79	0.0878	0.8816	79.68	0.0029	0.9992

Evaluation of Thermodynamic Parameters

In order to study the feasibility of the process and the potential application of the present adsorbent, the thermodynamic parameters are evaluated [57]. Thermodynamic parameters change in free energy ($\Delta G°$), enthalpy ($\Delta H°$) and entropy ($\Delta S°$) were determined in the temperature range of 303–333 K at an initial PS concentration of 200 mg/L and adsorbent dosage of 1.50 g/L. The thermodynamic parameters of the adsorption were calculated using Eq. (9) and the Van't Hoff (Eq. (10)):

$$\Delta G° = -RT \ln K_p \tag{9}$$

$$\ln K_p = \frac{\Delta S°}{R} - \frac{\Delta H°}{RT} \tag{10}$$

where K_p is the adsorption equilibrium constant (from Langmuir model, $K_p = K_L q_m$), R is the gas constant (8.314 J/mol K) and T is the adsorption temperature (K). A plot of $\ln K_p$ as a function of $10^3/T$ yielded a straight line (Fig. 10). The values of $\Delta H°$ and $\Delta S°$ obtained from the slope and intercept of the plot are summarized in Table 4. In general, the change of standard free energy ($\Delta G°$) for physisorption is in a range of −20 to 0 kJ/mol, and the chemisorption varies between −80 and −400 kJ/mol [58], [59] and [9]. The overall ΔG° of the adsorbent is negative values from −6.98 to −6.11 kJ/mol at the temperature range studied. The negative values for the standard free energy change, ΔG° showed the adsorption process for the FGHPs was feasible and spontaneous thermodynamically. However, the decrease in ΔG° absolute values with increase in temperature showed that the adsorption was not favorable at high temperatures[9] and [60]. This is confirmed by negative value of ΔH°, which revealed that the adsorption is exothermic and likely to be dominated by physical process in nature [58]. The negative values of ΔS° suggested a decrease in randomness at their solid–liquid interface and no significant changes occur in the internal structure of the adsorbent through the adsorption [54], [55] and [56].

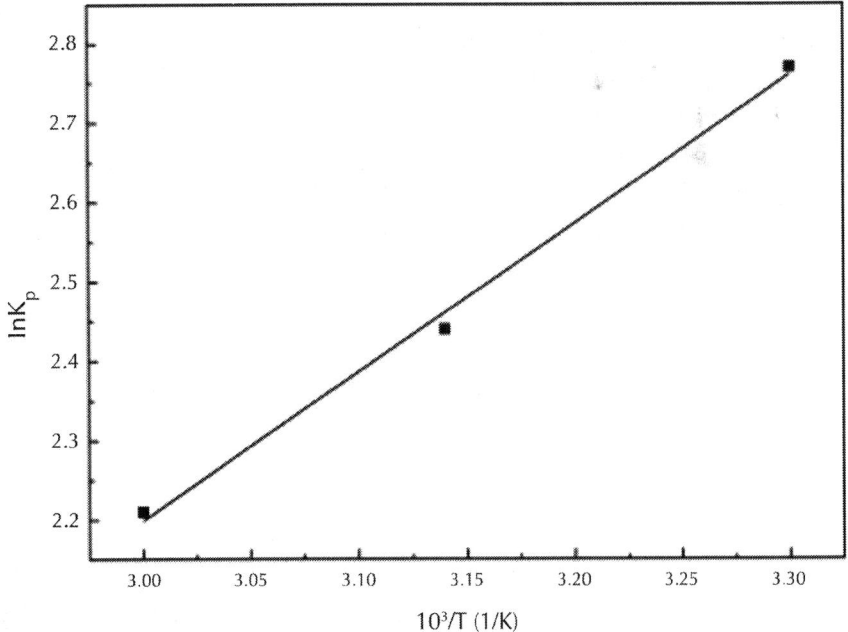

Figure 10: Van't Hoff plot for adsorption of PS by FGHPs.

Table 4: Thermodynamic parameters for adsorption of PS by FGHPs

Temperature (K)	$\Delta G°$ (kJ/mol)	$\Delta H°$ (kJ/mol)	$\Delta S°$ (J/mol K)
303	−6.98		
318	−6.46	−15.55	−28.43
333	−6.11		

Determination of Heavy Metals Leaching

The ingredients for the ammonia–soda process are salt brine (mainly from the sea) and limestone (from mines). They may have minor potential environmental impact. White mud, which is the residual fraction of sea water after extraction of sodium, is a less toxicity material [23] and [61]. The X-ray fluorescence spectra also showed that the white mud contains lower levels of heavy metals (Table 1). However, it is known that by-products of industrial waste materials as adsorbent may be more harmful than the original substance. Also, there is always the question of the fate of the adsorbed contaminants and how to prevent re-contamination of water sources. Since the potentially re-contamination in this work was heavy metals originate from sea water. Thus, the leaching of heavy metal ions in the treated water was measured by ICP spectroscopy, as reported in Table 5. After the adsorption process by 1.50 g/L of LSWM in aqueous system (pH 12.0), the concentration of heavy metals in the water was less than the standard limitation of effluent, meaning that the treated water can be safely discharged.

Table 5: Concentration of heavy metal ions in treated water after adsorption by 1.50 g/L of LSWM in aqueous system

Heavy metal ions	Cd	Cr	Cu	Pb	Si	Zn	Ni
Concentration (mg/L)	0.007	0.102	0.650	0.213	0.120	0.025	Not detected
Standard (mg/L)	0.1	1.5	5.0	1.0	10	2	1.0

CONCLUSIONS

Experimental data showed that FGHPs can remove PS effectively from aqueous solution. FGHPs displayed excellent removal efficiency at pH 12.0. The maximum capacities of FGHPs for PS were reached within 60 s. The removal efficiency increased significantly with increasing solution pH, and then decreased at solution pH above 12.0. The experimental equilibrium data can be fitted well by the Langmuir isotherm. The kinetic data were presented by the pseudo-second-order kinetic model. The adsorption process is exothermic in nature, and so a lower temperature is favorable. The maximum adsorptive capacity of FGHPs for PS was 3798.06 mg/g at 303 K. Therefore, FGHPs can be used to treat surfactant-containing wastewater effectively. The adsorption mechanism is co-precipitation/adsorption process that involves a combination of electrostatic attraction and hydrogen bonding. Liquid bridges through hydrogen bonding, adhesive and cohesive forces are linked up with the hydrate $MOH^+/M(OH)_2$ particles and surfactant molecules to form a three-dimensional network structure and lead to deposition. Therefore, both adhesion and cohesion play important role in controlling co-precipitation/adsorption process.

ACKNOWLEDGMENTS

This work was financially supported by the National Natural Science Foundation of China (No. 51178253) and the Natural Science Foundation of Shandong province in China (ZR2011EEM002).

REFERENCES

1. E. Ayranci, O. Duman, Removal of anionic surfactants from aqueous solutions by adsorption onto high area activated carbon cloth studied by in situ UV spectroscopy, J. Hazard. Mater. 148 (2007) 75–82.

2. X. Huang, T. Wu, Y. Li, D. Sun, G. Zhang, Y. Wang, G. Wang, M. Zhang, Removal of petroleum sulfonate from aqueous solutions using freshly generated magnesium hydroxide, J. Hazard. Mater. 219 (2012) 82–88.
3. H. Koyuncu, N. Yıldız, U. Salgın, F. Körog˘lu, A. Çalımlı, Adsorption of o-, m- and p-nitrophenols onto organically modified bentonites, J. Hazard. Mater. 185 (2011) 1332–1339.
4. J. Zhang, L. Zhang, J. Tang, L. Jiang, Interactions between poly (acrylamide) and surfactants of different headgroup charge, Colloids Surf. A 88 (1994) 33–39.
5. G. Zhang, X. Li, Y. Li, T. Wu, D. Sun, F. Lu, Removal of anionic dyes from aqueous solution by leaching solutions of white mud, Desalination 274 (2011) 255–261.
6. M.C. Biesinger, B.R. Hart, R. Polack, B.A. Kobe, R.S.C. Smart, Analysis of mineral surface chemistry in flotation separation using imaging XPS, Miner. Eng. 20 (2007) 152–162.
7. A.H. Mollah, C.W. Robinson, Pentachlorophenol adsorption and desorption characteristics of granular activated carbon—I. Isotherms, Water Res. 30 (1996) 2901–2906.
8. P. Waranusantigul, P. Pokethitiyook, M. Kruatrachue, E. Upatham, Kinetics of basic dye (methylene blue) biosorption by giant duckweed (Spirodela polyrrhiza), Environ. Pollut. 125 (2003) 385–392.
9. A. Li, Q. Zhang, G. Zhang, J. Chen, Z. Fei, F. Liu, Adsorption of phenolic compounds from aqueous solutions by a water-compatible hypercrosslinked polymeric adsorbent, Chemosphere 47 (2002) 981–989.
10. K.V. Kumar, S. Sivanesan, V. Ramamurthi, Adsorption of malachite green onto Pithophora sp., a fresh water algae: equilibrium and kinetic modelling, Process Biochem. 40 (2005) 2865–2872.
11. Q. Li, Q.-Y. Yue, Y. Su, B.-Y. Gao, J. Li, Two-step kinetic study on the adsorption and desorption of reactive dyes at cationic polymer/bentonite, J. Hazard. Mater. 165 (2009) 1170–1178.

12. X.-J. Wang, Y. Song, J.-S. Mai, Combined Fenton oxidation and aerobic biological processes for treating a surfactant wastewater containing abundant sulfate, J. Hazard. Mater. 160 (2008) 344–348.
13. I.D. Mall, V.C. Srivastava, N.K. Agarwal, Removal of orange-G and methyl violet dyes by adsorption onto bagasse fly ash—kinetic study and equilibrium isotherm analyses, Dyes Pigments 69 (2006) 210–223.
14. A. Fakhru'l-Razi, A. Pendashteh, L.C. Abdullah, D.R.A. Biak, S.S. Madaeni, Z.Z. Abidin, Review of technologies for oil and gas produced water treatment, J. Hazard. Mater. 170 (2009) 530–551.
15. Z.-L. Ye, S.-H. Chen, S.-M. Wang, L.-F. Lin, Y.-J. Yan, Z.-J. Zhang, J.-S. Chen, Phosphorus recovery from synthetic swine wastewater by chemical precipitation using response surface methodology, J. Hazard. Mater. 176 (2010) 1083–1088.
16. Y. Song, P. Yuan, B. Zheng, J. Peng, F. Yuan, Y. Gao, Nutrients removal and recovery by crystallization of magnesium ammonium phosphate from synthetic swine wastewater, Chemosphere 69 (2007) 319–324.
17. J.-M. Sun, S.-Y. Chang, R. Li, J.-C. Huang, Factors affecting co-removal of chromium through copper precipitation, Sep. Purif. Technol. 56 (2007) 57–62.
18. J. Leentvaar, M. Rebhun, Effect of magnesium and calcium precipitation on coagulation–flocculation with lime, Water Res. 16 (1982) 655–662.
19. S. Netpradit, P. Thiravetyan, S. Towprayoon, Application of 'waste' metal hydroxide sludge for adsorption of azo reactive dyes, Water Res. 37 (2003) 763–772.
20. V. Vimonses, B. Jin, C.W. Chow, C. Saint, Enhancing removal efficiency of anionic dye by combination and calcination of clay materials and calcium hydroxide, J. Hazard. Mater. 171 (2009) 941–947.
21. C. Namasivayam, S. Sumithra, Removal of direct red 12B and methylene blue from water by adsorption onto Fe (III)/Cr(III)

hydroxide, an industrial solid waste, J. Environ. Manag. 74 (2005) 207–215.
22. M.-X. Zhu, L. Lee, H.-H. Wang, Z. Wang, Removal of an anionic dye by adsorption/precipitation processes using alkaline white mud, J. Hazard. Mater. 149 (2007) 735–741.
23. L. Shi, H.-J. Luo, Preparation of soil nutrient amendment using white mud produced in ammonia–soda process and its environmental assessment, Trans. Nonferrous Metals Soc. China 19 (2009) 1383–1388.
24. T. Wu, D.J. Sun, Y.J. Li, G.C. Zhang, Treatment of waste cutting oil emulsions by leaching solutions of white mud, Adv. Mater. Res. 356 (2012) 1570–1574.
25. U.F. Alkaram, A.A. Mukhlis, A.H. Al-Dujaili, The removal of phenol from aqueous solutions by adsorption using surfactant-modified bentonite and kaolinite, J. Hazard. Mater. 169 (2009) 324–332.
26. C. Zhang, S. Yang, H. Chen, H. He, C. Sun, Adsorption behavior and mechanism of reactive brilliant red X-3B in aqueous solution over three kinds of hydrotalcite-like LDHs, Appl. Surf. Sci. 301 (2014) 329–337.
27. M. Özacar, C. Soykan, I.A. Sengi_l, Studies on synthesis, characterization, and metal adsorption of mimosa and valonia tannin resins, J. Appl. Polym. Sci. 102 (2006) 786–797.
28. G. Carja, L. Dartu, K. Okada, E. Fortunato, Nanoparticles of copper oxide on layered double hydroxides and the derived solid solutions as wide spectrum active nano-photocatalysts, Chem. Eng. J. 222 (2013) 60–66.
29. F. Cavani, F. Trifirò, A. Vaccari, Hydrotalcite-type anionic clays: preparation, properties and applications, Catal. Today 11 (1991) 173–301.
30. Y. Zhao, H. Ding, Q. Zhong, Preparation and characterization of aminated graphite oxide for CO_2 capture, Appl. Surf. Sci. 258 (2012) 4301–4307.

31. Y. Park, G.A. Ayoko, R.L. Frost, Characterisation of organoclays and adsorption of p-nitrophenol: environmental application, J. Colloid Interface Sci. 360 (2011) 440–456.
32. B. Nandi, A. Goswami, M. Purkait, Removal of cationic dyes from aqueous solutions by kaolin: kinetic and equilibrium studies, Appl. Clay Sci. 42 (2009) 583–590.
33. S. Tahir, N. Rauf, Removal of a cationic dye from aqueous solutions by adsorption onto bentonite clay, Chemosphere 63 (2006) 1842–1848.
34. M. Alkan, Ö. Demirbas, M. Doğan, Electrokinetic properties of kaolinite in mono- and multivalent electrolyte solutions, Micropor. Mesopor. Mater. 83 (2005) 51–59.
35. Z.M. Aman, K. Olcott, K. Pfeiffer, E.D. Sloan, A.K. Sum, C.A. Koh, Surfactant adsorption and interfacial tension investigations on cyclopentane hydrate, Langmuir 29 (2013) 2676–2682.
36. Z.M. Aman, E.P. Brown, E.D. Sloan, A.K. Sum, C.A. Koh, Interfacial mechanisms governing cyclopentane clathrate hydrate adhesion/cohesion, Phys. Chem. Chem. Phys. 13 (2011) 19796–19806.
37. P. Wang, C. Li, H. Gong, H. Wang, J. Liu, Morphology control and growth mechanism of magnesium hydroxide nanoparticles via a simple wet precipitation method, Ceram. Int. 37 (2011) 3365–3370.
38. Y. Liu, Q. Li, S. Gao, J.K. Shang, Exceptional As(III) sorption capacity by highly porous magnesium oxide nanoflakes made from hydrothermal synthesis, J. Am. Ceram. Soc. 94 (2011) 217–223.
39. E. Eren, A. Tabak, B. Eren, Performance of magnesium oxide-coated bentonite in removal process of copper ions from aqueous solution, Desalination 257 (2010) 163–169.
40. X. Pan, Y. Wang, Z. Chen, D. Pan, Y. Cheng, Z. Liu, Z. Lin, X. Guan, Investigation of antibacterial activity and related mechanism of a series of nano-Mg(OH)2, ACS Appl. Mater. Interfaces 5 (2013) 1137–1142.

41. Y. Li, B. Gao, T. Wu, B. Wang, X. Li, Adsorption properties of aluminum magnesium mixed hydroxide for the model anionic dye reactive Brilliant Red K-2BP, J. Hazard. Mater. 164 (2009) 1098–1104.
42. M. Doğan, Y. Özdemir, M. Alkan, Adsorption kinetics and mechanism of cationic methyl violet and methylene blue dyes onto sepiolite, Dyes Pigments 75 (2007) 701–713.
43. S. Wang, H. Li, Dye adsorption on unburned carbon: kinetics and equilibrium, J. Hazard. Mater. 126 (2005) 71–77.
44. A. De Lazzer, M. Dreyer, H. Rath, Particle–surface capillary forces, Langmuir 15 (1999) 4551–4559.
45. Z.M. Aman, E.D. Sloan, A.K. Sum, C.A. Koh, Lowering of clathrate hydrate cohesive forces by surface active carboxylic acids, Energy Fuels 26 (2012) 5102–5108.
46. G. Aspenes, L. Dieker, Z. Aman, S. Høiland, A. Sum, C. Koh, E. Sloan, Adhesion force between cyclopentane hydrates and solid surface materials, J. Colloid Interface Sci. 343 (2010) 529–536.
47. Y.I. Rabinovich, M.S. Esayanur, K.D. Johanson, J.J. Adler, B.M. Moudgil, Measurement of oil-mediated particle adhesion to a silica substrate by atomic force microscopy, J. Adhes. Sci. Technol. 16 (2002) 887–903.
48. S. Biggs, R.G. Cain, R.R. Dagastine, N.W. Page, Direct measurements of the adhesion between a glass particle and a glass surface in a humid atmosphere, J. Adhes. Sci. Technol. 16 (2002) 869–885.
49. Y.I. Rabinovich, M.S. Esayanur, B.M. Moudgil, Capillary forces between two spheres with a fixed volume liquid bridge: theory and experiment, Langmuir 21 (2005) 10992–10997.
50. M. Özacar, I.A. Sengil, Adsorption of reactive dyes on calcined alunite from aqueous solutions, J. Hazard. Mater. 98 (2003) 211–224.
51. V.K. Gupta, S.K. Srivastava, D. Mohan, Equilibrium uptake, sorption dynamics, process optimization, and column operations for the removal and recovery of malachite green

from wastewater using activated carbon and activated slag, Ind. Eng. Chem. Res. 36 (1997) 2207–2218.
52. K.P. Singh, D. Mohan, S. Sinha, G. Tondon, D. Gosh, Color removal from wastewater using low-cost activated carbon derived from agricultural waste material, Ind. Eng. Chem. Res. 42 (2003) 1965–1976.
53. M. Barhoumi, I. Beurroies, R. Denoyel, H. Saïd, K. Hanna, Coadsorption of alkylphenols and nonionic surfactants onto kaolinite, Colloids Surf. A 219 (2003) 25–33.
54. R. Garcia-Delgado, L. Cotoruelo-Minguez, J. Rodriguez, Equilibrium study of single-solute adsorption of anionic surfactants with polymeric XAD resins, Sep. Sci. Technol. 27 (1992) 975–987.
55. E. Zong, D. Wei, H. Wan, S. Zheng, Z. Xu, D. Zhu, Adsorptive removal of phosphate ions from aqueous solution using zirconia-functionalized graphite oxide, Chem. Eng. J. 221 (2013) 193–203.
56. A. Viinikanoja, J. Kauppila, P. Damlin, E. Mäkilä, J. Leiro, T. Ääritalo, J. Lukkari, Interactions between graphene sheets and ionic molecules used for the shearassisted exfoliation of natural graphite, Carbon 68 (2014) 195–209.
57. V. Vimonses, S. Lei, B. Jin, C.W. Chow, C. Saint, Kinetic study and equilibrium isotherm analysis of Congo Red adsorption by clay materials, Chem. Eng. J. 148 (2009) 354–364.
58. B. Chu, B. Baharin, Y. Che Man, S. Quek, Separation of vitamin E from palm fatty acid distillate using silica: I. Equilibrium of batch adsorption, J. Food Eng. 62 (2004) 97–103.
59. Y. Ho, G. McKay, Comparative sorption kinetic studies of dye and aromatic compounds onto fly ash, J. Environ. Sci. Heal. A 34 (1999) 1179–1204.
60. A.R. Kul, H. Koyuncu, Adsorption of Pb(II) ions from aqueous solution by native and activated bentonite: kinetic, equilibrium and thermodynamic study, J. Hazard. Mater. 179 (2010) 332–339.

61. CEFIC-ESAPA, IPPCB BAT Refence Document Large Volume Solid Inorganic Chemicals Family, Process BREF for Soda Ash, Issue No: 3, 2004.

Electrochemical Treatment of Effluents from Petroleum Industry Using a Ti/RuO$_2$ Anode

Iranildes D. Santos[a], Márcia Dezotti[b], and Achilles J.B. Dutra[a]

[a]Metallurgical and Materials Engineering Program, Federal University of Rio de Janeiro, Rio de Janeiro, Brazil

[b]Chemical Engineering Program, Federal University of Rio de Janeiro, Rio de Janeiro, Brazil

ABSTRACT

In this paper the efficiency of Ti/RuO$_2$ anode in degrading organic substances, present in wastewaters from petroleum industry, before their discharge or reuse was investigated. Results indicated that the Ti/RuO$_2$ electrode can be an efficient alternative for treatment of those effluents. The COD (Chemical Oxygen Demand) removals, after 120 min of electrolysis, with a current density of 10 mA cm^{-2}, anodic area of 107 cm^2, flow rate of 0.54 mL s^{-1} and at 25 °C, were above 96%, for effluent AF (After Flotation), with 712 mg L^{-1} COD, and 87% for effluent BF (Before Flotation), with 833 mg L^{-1} COD. An almost complete COD removal from both effluents was achieved when current density was increased from 10 to 30 mA cm^{-2} with anodic area of 107 cm^2, flow rate of 0.54 mL s^{-1} and at 25 °C. The increase of current density also favored a decrease of the electrolysis time necessary to achieve a complete COD removal from both effluents, BF and AF. However, current density increase also led to a higher specific energy consumption. For effluent BF, the cost of the energy necessary to achieve a complete COD removal in 60 min with a current density of 30 mA cm^{-2} with anodic area of 107 cm^2, flow rate of 0.54 mL s^{-1} and at 25 °C was around US$ 38/kg COD, while for effluent AF, the energy cost for total COD removal after 30 min of electrolysis at a current density of 30 mA cm^{-2} was only US$ 28/kg COD.

INTRODUCTION

In recent years there has been a significant increase of the volume of effluents generated during the operations of exploitation, production and refining of oil and gas in the world. In 2009, Brazilian refineries consumed 254,093 m^3 of water per day [1] and in the oil production and exploration, the effluent volume may be as high as 10 times the volume of oil generated [2].

In general, the effluent generated in the exploitation, production and refining of oil contains high concentrations of organic and inorganic pollutants, besides chloride (which can vary from 15,000 to 300,000 mg L^{-1}), suspended solids, oil, grease and chemical additives, such as corrosion inhibitors and surfactants that are added during oil production operations [3], [4] and [5]. Consequently, these effluents need to be treated prior to discharge or reuse.

There are a number of different technologies for oily wastewater treatment, such as flotation [4], [6] and [7], coagulation and flocculation [8] and [9], advanced oxidation processes [10], [11] and [12] and biological processes [13] and [14]. However the choice of the technology to be used for effluent treatment depends, among other factors, on the effluent composition and the final destination of the effluent (discharge or reuse).

The electrochemical technology, using different electrode materials, has proved to be both an efficient and viable alternative for the treatment of various types of effluents [4], [15], [16], [17], [18], [19], [20],[21] and [22]. For the electrochemical treatment of produced water, Ma and Wang [23], used a metal/graphite/iron (MCFe) electrode and obtained COD and suspended solids removals of respectively 90% and 99%, after 6 min of electrolysis. Santos et al. [24] used a Ti/Ru$_{0.34}$Ti$_{0.66}$O$_2$ electrode for the produced water treatment from an oil/water separator and obtained a 57% COD removal after 70 h of electrolysis at 50 °C with a current density of 100 mA cm^{-2}. In general, the main restrictions to the use of electrochemical processes for the treatment of saline effluents include the low corrosion resistance of electrode materials and the formation of organochlorine compounds during electrolysis [25] and [26].

As the Ti/RuO$_2$ electrode has already been successfully used in saline environments, such as in the chlor-alkali industry [3], the objective of this study was evaluating its use in the degradation of organic substances present in effluents from the petroleum industry prior discharge or reuse, in order to improve the quality of the effluent treated by flotation.

MATERIALS AND METHODS

Effluent Source

The effluents samples were collected from the Ferry Terminal Almirante Barroso (TEBAR) located in São Sebastião/SP/Brazil. This terminal is responsible for the movement of 4 billion gallons oil per month, representing 50% of the oil consumed by the country. The samples were collected before (BF) and after (AF) the flotation unit.

Electrochemical Behavior of Ti/RuO$_2$ Electrode

The cyclic voltammograms of the Ti/RuO$_2$ electrode in the effluents BF and AF were plotted using an Ivium model Compactstat-e1030 potentiostat/galvanostat. For these tests a 25 mL cell was used. The working electrode was a Ti/RuO$_2$ plate with an effective exposed area of 1.35 cm^2; the counter electrode was a platinum wire and the Ag/AgCl electrode, saturated in KCl ($E° = 0.198$ V vs. SHE), was used as reference. All the tests were carried out at room temperature (around 25 °C) without stirring.

The voltammetric tests were carried out in NaCl solutions (with concentrations of 51 g L^{-1}) in order to simulate effluent salinity, and with real effluents BF and AF.

Electrochemical Tests and Chemical Analysis

The electrolytic cell used for electrooxidation of the organic compounds in the effluent was a plug-flow reactor (PFR), operated in a semi-continuous mode, as shown in Fig. 1. Titanium and Ti/RuO$_2$ (provided by De Nora®) plates were used as cathode and anode, respectively. The anode potential and pH were measured during the electrolysis; the current densities tested were 10 and 30 mA cm^{-2}. All the experiments were carried out with 400 mL of wastewater and a flow rate 0.54 mL min^{-1}, rendering a retention time of 12.3 min. The COD was determined for different electrolysis times by the dichromate colorimetric method according to Freire and Sant'anna [27] which is a specific method for saline effluents.

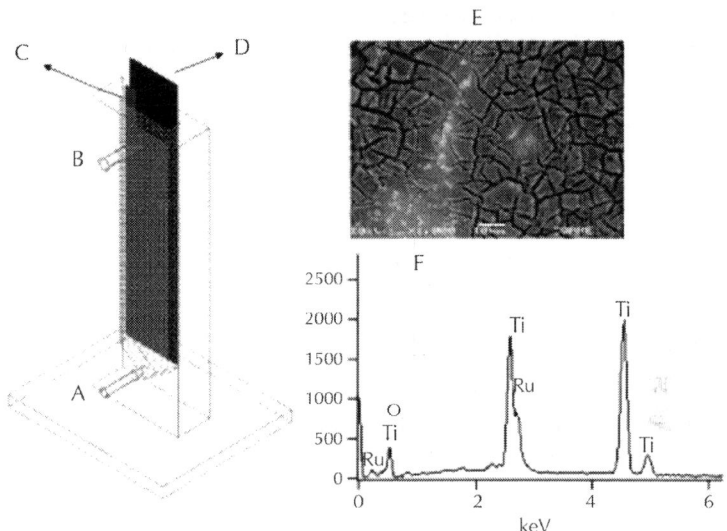

Figure 1: Scheme of the electrolytic cell used in the experiments. (A) Effluent entrance; (B) effluent exit; (C) cathode: titanium plate; (D) anode: Ti/RuO$_2$ plate; (E) SEM micrographs of Ti/RuO$_2$ anode and (F) EDS spectrum of Ti/RuO$_2$ anode.

A Shimadzu, 1601PC model UV/Vis spectrophotometer was used to monitor organic compounds degradation [15] and [28].

Energy dispersive X-ray fluorescence (EDXRF) was used to identify and quantify the elements adsorbed on the cathode surface after electrolysis.

Before each electrolysis test, the residual materials adsorbed on the cathode surface were anodically stripped by electrolysis using a 2 mol L^{-1} HCl solution with a current density of 1 A cm^{-2} for 2 min.

The energy consumption for the removal of 1 kg of COD was obtained by Eq. (1) and the energy cost for the removal of 1 kg COD from the effluents BF and AF by Eq. (2), considering the industrial kWh price provided by Light S.A. (local electricity distribution company), in Rio de Janeiro State, Brazil, in October 2012.

$$EC \equiv \frac{UIt/V}{\Delta COD} \times 10^3$$

(1)

where EC is the energy consumption for the removal of 1 kg of COD; t is the electrolysis time (h), U is the cell voltage (V), I is the applied current (A), V is the solution volume (L), and COD is the difference of COD (mg L^{-1}) before and after the electrolysis.

$$C_c = C_{EC} \times P_{kWh} \tag{2}$$

where C_c is the energy cost for removal of 1 kg COD (US$/kg$_{COD}$); C_{EC} the energy consumption (kW h/kg$_{COD}$); P_{kWh} the industrial kW h price, in October/2012 (US$ 0.98/kW h) for consumptions higher than 300 kW h.

RESULTS AND DISCUSSION

Effluent Characterization

The initial characterization of effluents BF and AF is presented in Table 1. It can be observed that both effluents presented a large amount of NaCl, rendering a high electrical conductivity. The lower COD found in effluent AF is due to partial organic matter removal by the flotation process [8] and [24].

Table 1: Typical characteristics of effluents from TEBAR, collected before (BF) and after (AF) the flotation unit

Parameter	Effluent BF	Effluent AF
pH	7.2 ± 0.03	7.3 ± 0.05
Conductivity (mS cm^{-1})	0.2 ± 0.02	0.2 ± 0.02
Chloride (g L^{-1})	51.4 ± 0.8	48.8 ± 0.39
COD (mg L^{-1})	833.3 ± 64.3	712.1 ± 64.3

Voltammetric Behavior of the Ti/RuO$_2$ Electrode

The voltammetric behavior of the Ti/RuO$_2$ electrode in a synthetic aqueous solution with 51 g L^{-1} of Cl$^-$ions, as NaCl (approximately the same concentration of Cl-in the wastewater), and in the presence of effluents BF and AF is presented in Fig. 2. It can be observed that the voltammograms shapes for the NaCl solution and effluent BF were very close, however, for potential values higher than 0.65 V, higher current density values were obtained for the voltammogram with the NaCl solution alone, probably due to the absence of organic compounds, since their presence on the electrode surface reduces its electrocatalytic activity. Compared to the saline solution voltammogram, a decrease in current density was also observed in the presence of both effluents BF and AF in the potential range between 0.8 and 1.5 V. This behavior can be attributed to the adsorption of intermediate organic products (such as chlorophenols, which are more resistant to degradation) and to the oxidation of Cl$^-$ ions to Cl$_2$ during the electrolysis in chloride-containing medium [18], [25], [29], [30], [31] and [32]. Anodic and cathodic peaks were observed only in effluent AF. For effluent BF, only an attenuated wave was observed at 0.95 V; at the same potential an evident peak (shoulder) was observed for the effluent AF. The absence of peaks for effluent BF, collected before the flotation unit, can be attributed to the intense anode passivation, which causes a decrease on the current density and inhibits peaks formation. The anodic peaks observed for effluent AF can be attributed to the oxidation reaction of organic compounds, while the cathodic peak suggests the reduction of adsorbed intermediate organic products previously formed on the electrode surface during the anodic scan [18],[29] and [33].

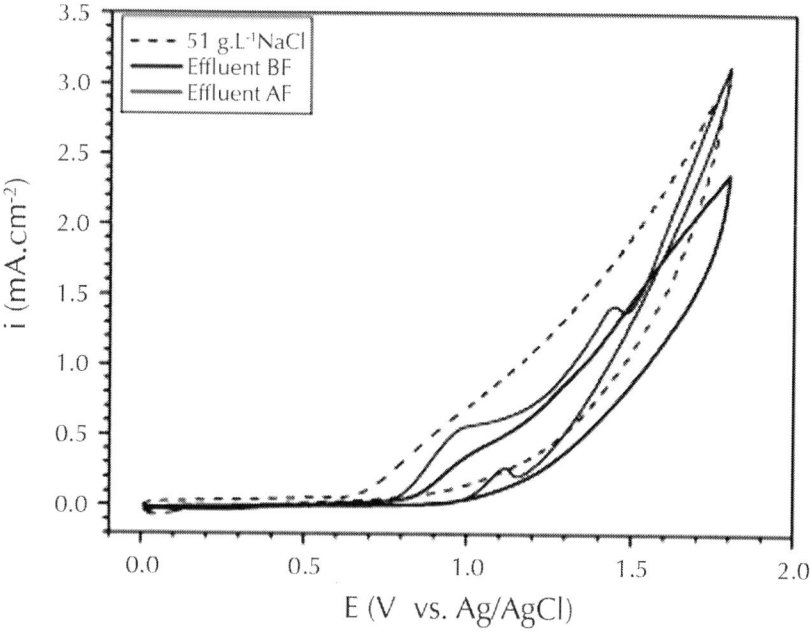

Figure 2: Cyclic voltammograms performed at Ti/RuO$_2$ anode in 51 g L^{-1} NaCl solution and effluents BF and AF. Anodic area: 1.35 cm^2, T: 25 °C and scan rate: 0.02 V s^{-1}.

Depending on the electrooxidation conditions [19] and [34] the adsorption of organic matter on the electrode surface can reduce the efficiency of the Ti/RuO$_2$ electrode for the electrooxidation of organic compounds. The behavior of Ti/RuO$_2$ electrode in the presence of effluents BF and AF, before and after reactivation at 1.8 V during 60 s is presented in Fig. 3. It can be observed that after reactivation the adsorbed blocking layer on the electrode surface (according to Fig. 2) was removed, indicating that the catalytic properties of the electrode were reestablished. This behavior was similar for both effluents. The recovery of the electrode catalytic properties can be attributed to the oxidation of the organic matter layer, which was blocking the anode surface, by the action of oxidizing agents generated during electrolysis, such as Cl$_2$ and/or OCl$^-$, cleaning up the electrode surface.

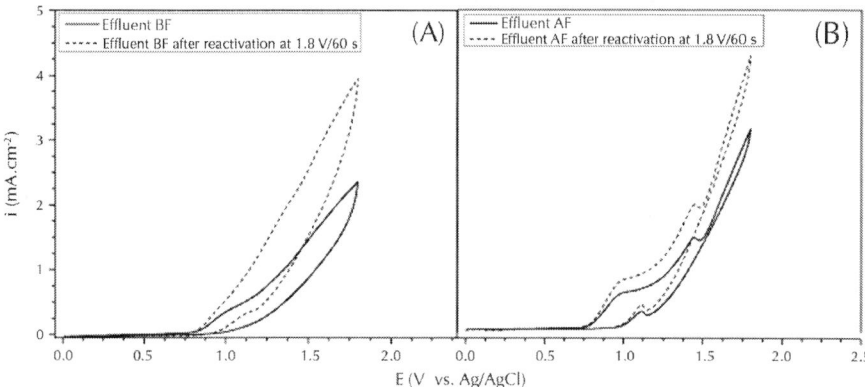

Figure 3: Cyclic voltammograms performed with a Ti/RuO$_2$ electrode in the presence of effluents BF (A) and AF (B), before and after reactivation at 1.8 V during 60 s. Anodic area: 1.35 cm^2, T: 25 °C and scan rate: 0.02 V s^{-1}.

Performance of Electrochemical Oxidation

The influence of electrolysis time on COD removal with a current density of 10 mA cm^{-2} is presented in Fig. 4. It can be observed that the COD dropped throughout the electrolysis for both effluents. A higher COD removal, around 96%, was obtained after 120 min of electrolysis for effluent AF, while for effluent BF the removal was around 87% under the same experimental conditions. The lower COD removal observed for effluent BF, can be attributed to the slightly higher concentration of NaCl present in this effluent, allied to its larger organic load; under these conditions a higher concentration of oxidizing agent (OCl$^-$) was achieved, favoring the formation of chlorophenols which are more resistant to degradation Santos et al. [35] and [36].

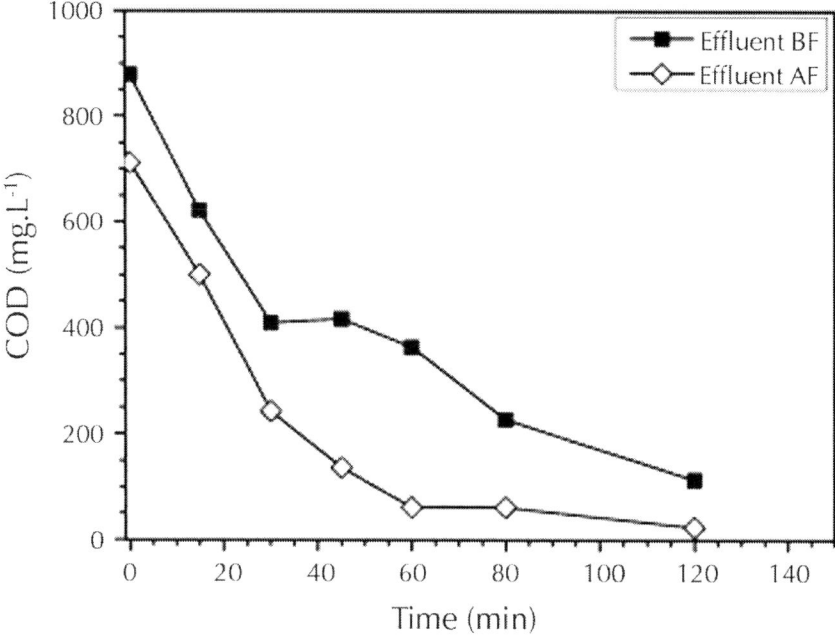

Figure 4: Influence of electrolysis time on the COD removal of effluents BF and AF. Anodic area: 107 cm², i = 10 mA cm^{-2}, effluent volume: 0.4 L, flow rate: 0.54 mL s^{-1}, T: 25 °C.

After 2 h of electrolysis with a current density of 10 mA cm^{-2}, a slight layer of adsorbed material was observed on the cathode surface, for both effluents. The chemical composition of the adsorbed material on the cathode surface after 2 h of electrolysis of effluent BF is presented in Table 2. The presence of different types of metallic hydroxides, identified after calcination as CaO and MgO can be observed. The formation of calcium and magnesium hydroxides can be attributed to the local pH increase near cathode surface, due to cathodic hydrogen evolution, according to reaction (3), which leads to the precipitation of metallic hydroxides. The presence of these metallic oxides/hydroxides was also reported by others researchers [37], [38] and [39].

$$2H_2O + 2e^- \rightarrow H_2 + 2OH^- \tag{3}$$

Table 2: Chemical composition of the material adsorbed on the cathode surface, after 2 h of electrolysis of effluent BF at 10 mA cm^{-2}

Compounds	Average (%)	Compounds	Average (%)
CaO	48.90	SO_3	0.37
MgO	44.64	ZrO_2	0.07
SiO_2	2.00	Br	0.06
P_2O_5	1.61	MnO	0.05
SrO	1.57	Yb_2O_3	0.04
BaO	0.47	CuO	0.03
Fe_2O_3	0.16	ZnO	0.01

The pH variation as a function of electrolysis time during the treatment of effluents BF and AF is presented in Fig. 5. It can be observed that the increase of electrolysis time leads to a decrease in the pH for time periods shorter than 30 min. This behavior can be attributed to the oxidation of organic compounds to acid compounds of low molecular weight, such as oxalic acid, which can be easily oxidized to CO_2 [17] . After 30 min of electrolysis, the pH starts to increase, reaching 7.9 after 120 min of electrolysis, due to the oxidation of chloride ions on the anode surface producing chlorine and/or hypochlorite which become the predominating species [18] and [37]. The predominance of chlorine or hypochlorite depends on the pH, since the ionization constant of hypochlorous acid is 2.95×10^{-8} (pK_a = 7.53).

The influence of electrolysis time on the COD removal from effluents BF and AF with different current densities is presented in Fig. 6. It can be observed that for both effluents, COD removal increases with the electrolysis time, independent of the current density applied.

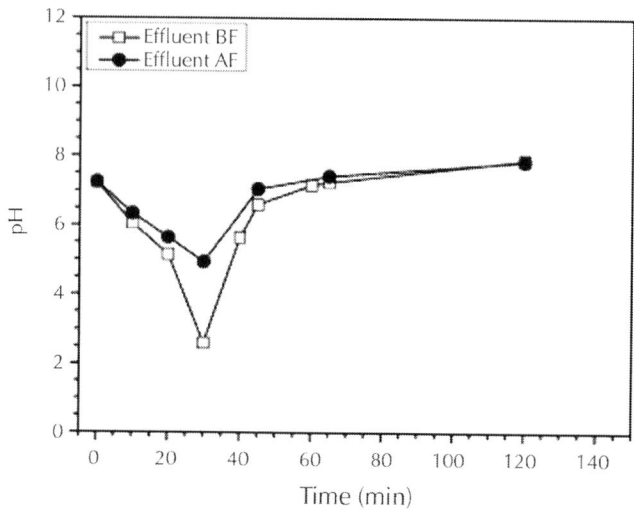

Figure 5: pH variation as a function of electrolysis time for effluents BF and AF. Anodic area: 107 cm², i = 10 mA cm^{-2}, flow rate: 0.54 mL s^{-1}, T: 25 °C.

However, a higher COD removal was obtained when the current density was increased from 10 to 30 mA cm^{-2}, indicating that, under the latter condition (30 mA cm^{-2}) a higher hypochlorite generation occurs, favoring organic compounds electrooxidation by the indirect mechanism (oxidation of organic compounds by oxidizing agents generated during the electrolysis). COD removals of 74% and 59% were obtained after 50 min of electrolysis with a current density of 10 mA cm^{-2} for the effluents BF and AF respectively. For a current density of 30 mA cm^{-2}, only 50 and 30 min were necessary for complete COD removal from effluents BF and AF, respectively. These results match fairly well with those obtained in a previous work [35] which indicated that the organochlorides formed during chloride ions reduction, through the reaction of chlorine/hypochloride with organics, are oxidized later, despite their resistance to oxidation.

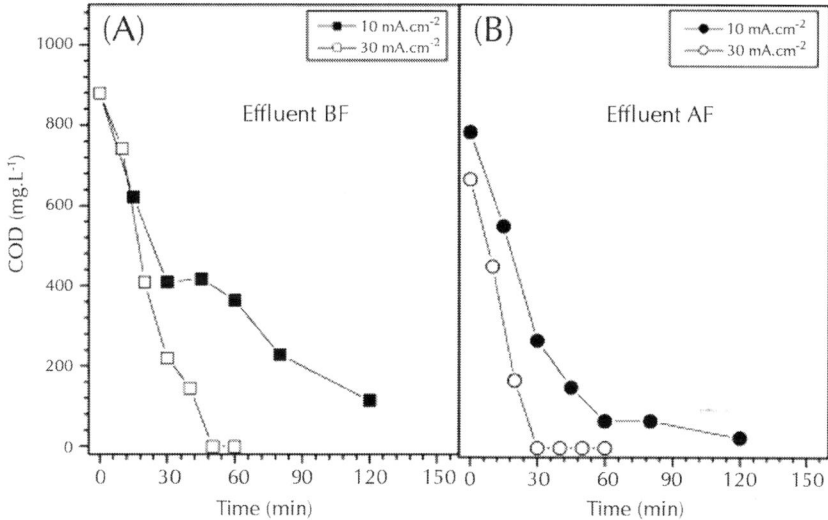

Figure 6: Influence of electrolysis time on COD removal from effluents BF and AF for different current densities. Anodic area: 107 cm², flow rate: 0.54 mL s^{-1}, T: 25 °C.

Specific Energy Consumption and Energy Cost

The influence of electrolysis time on the energy consumption required for the removal of 1 kg of COD (organic compounds including oils and greases) from effluents BF and AF with different current densities is shown in Fig. 7A and B, respectively. An increase in energy consumption throughout the electrolysis was observed for both effluents. A higher energy consumption for COD removal was observed when the current density was increased from 10 to 30 mA cm^{-2}. For effluent BF, after 60 min of electrolysis, specific energy consumptions of 18 and 40 kWh kg COD^{-1}, were observed for current densities of 10 and 30 mA cm^{-2} respectively. For effluent AF, under the same electrolysis conditions, specific energy consumptions of 14 and 50 kWh kg COD^{-1}, were observed for current densities of 10 and 30 mA cm^{-2} respectively. The higher energy consumption observed, for both effluents BF and AF, with a current density of 30 mA cm^{-2} can be attributed to excessive hypochlorite formation since Fig. 6 indicates that after 60 min of electrolysis with a current density of 30 mA cm^{-2} a complete COD removal for both effluents had already been achieved.

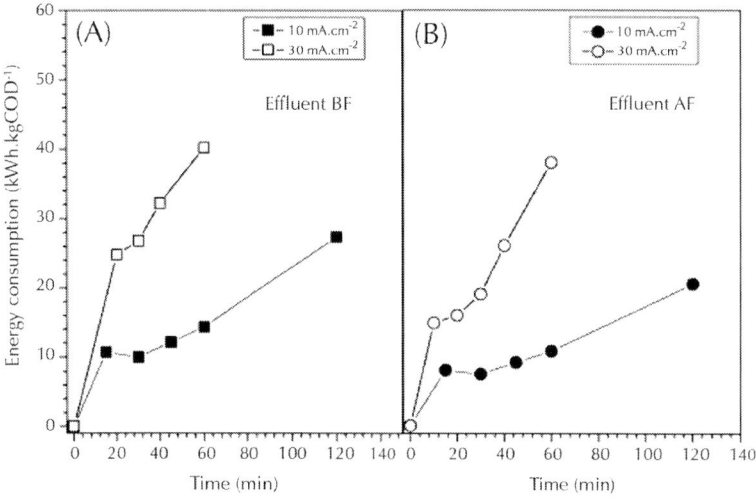

Figure 7: The influence of electrolysis time on the energy consumption for the removal of 1 kg of COD from effluents BF (A) and AF (B) at different current densities. Anodic area: 107 cm^2, flow rate: 0.54 mL s^{-1}, T: 25 °C.

The influence of electrolysis time on energy consumption and COD amount from effluents BF and AF with a current density of 10 mA cm^{-2} is presented in the Fig. 8. It can be observed that for both effluents (Fig. 8A and B) the energy consumption increased with the decrease of COD and electrolysis time [40]. For effluent BF (Fig. 8A), the COD remained constant during the time interval between 30 and 50 min. This behavior is, probably, due to the formation of organochlorine compounds, which are more difficult to degrade [35]. A higher specific energy consumption was observed for effluent AF when compared with effluent BF, probably due to a more intense hypochlorite formation since the effluent AF presents a smaller COD content (Table 1). After 60 min of electrolysis with a current density of 10 mA cm^{-2} a specific energy consumption of 0.026 kW h mg^{-1} was achieved for the removal of 58.7% (206.1 mg) of the COD from effluent BF. While for the effluent AF, after 30 min of electrolysis with a current density of 10 mA cm^{-2} a specific energy consumption of 0.19 kW h mg^{-1} was achieved for the removal of 65.9% (187.9 mg) of the COD.

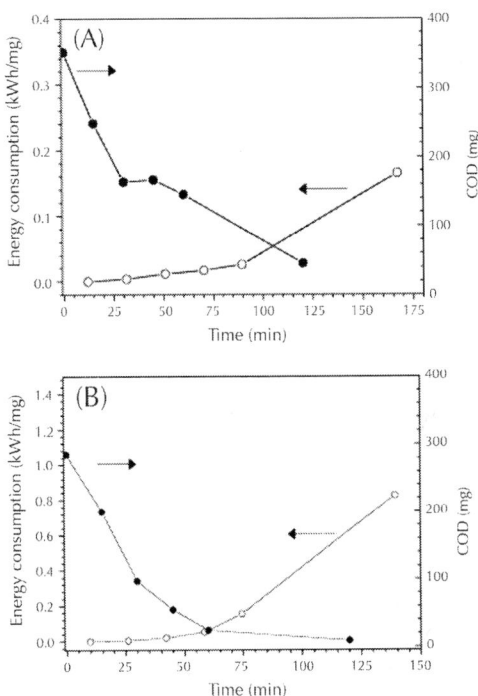

Figure 8: Influence of electrolysis time on energy consumption and COD removal from 400 mL of effluents BF (A) and AF (B). Anodic area: 107 cm^2, current density: 10 mA cm^{-2}, flow rate: 0.54 mL s^{-1}, T: 25 °C.

The absorbance spectra of effluent BF before and after different electrolysis times are presented in Fig. 9. It can be observed that the absorbance after 30 min was significantly reduced when compared to the untreated effluent throughout the entire range of wavelength studied, indicating that most part of the organic compounds were oxidized. The presence of an absorption band between 250 and

330 nm was observed after 60 min of electrolysis. This band may indicate the presence of both organochloride compounds and hypochlorite, which absorb in this band range, according to the insert of Fig. 9, which confirms hypochlorite absorbance band in this range. The interference of hypochlorite in the treated effluent can be eliminated by the addition of a reducing agent, such as sodium bisulfate, according to Santos et al. [35], leaving in the solution only the organic compounds. Then, the higher energy consumption observed in Fig. 8, for both effluents, can be attributed to hypochlorite formation in excess, and not necessarily to the oxidation of organic compounds.

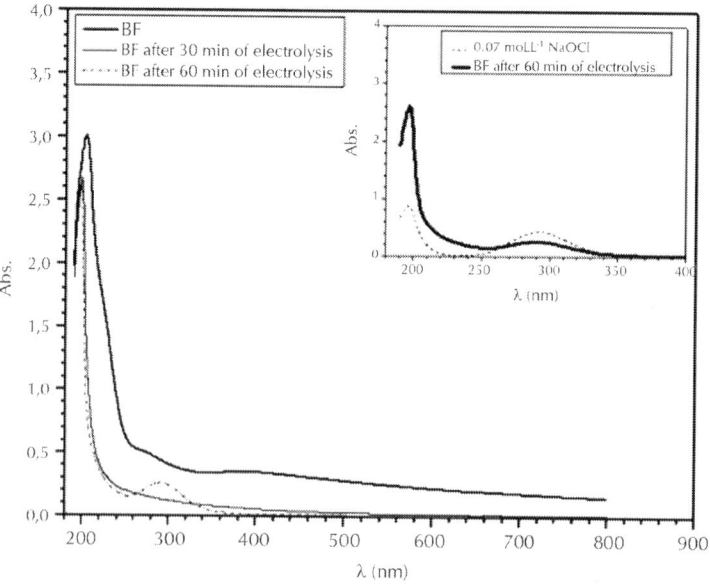

Figure 9: UV spectra of effluent BF before and after different electrolysis times. Anodic area: 107 cm², current density: 10 mA cm⁻², flow rate: 0.54 mL s⁻¹, T: 25 °C. Inset: UV spectra at 190–400 nm of effluent BF after 60 min of electrolysis and 0.07 mol L⁻¹ solution of sodium hypochlorite.

The electrolysis time, specific energy consumption and energy cost for COD removal from effluents BF and AF with different current densities for disposal according to the limits established by the environmental legislation of the State of Rio de Janeiro/Brazil for liquid effluent from industrial processes are shown in Table 3. It can be observed that the increase of current density led to an increase in the energy cost. On the other hand, it favored a decrease in the electrolysis time for COD removal from both effluents. For effluent BF, after 30 min of electrolysis with a current density of 30 mA cm^{-2} the energy cost necessary to remove 75% of COD was around US$ 26/kg COD, while for effluent AF, with the same current density, after 20 min of electrolysis the energy cost to remove 79% of COD was US$ 21/kg COD.

Table 3: Electrolysis time, specific energy consumption and energy cost for COD removal from effluents BF and AF with different current densities for disposal according to the limits established by the environmental legislation of the state of Rio de Janeiro/Brazil [41] for liquid effluents from industrial processes. Anodic area: 107 cm^2, flow rate: 0.54 mL s^{-1}, T: 25 °C

Parameters	Standard limits	$i = 10$ mA cm^{-2}		$i = 30$ mA cm^{-2}	
		BF	AF	BF	AF
Time (min)	–	120	30	30	20
COD (mg L^{-1})	250	113.6	242.4	219.7	151.5
Energy consumption (kW h kg COD^{-1})	–	23.1	10	26.8	21.3
Energy cost (US$. kg COD^{-1})	–	22.7	9.8	26.3	20.9

Initial concentration of COD: Effluent BF = 878.79 mg L^{-1} and effluent AF = 712.12 mg L^{-1}.

The electrolysis time, specific energy consumption and energy cost for the complete COD removal from effluents BF and AF with different current densities are presented in Table 4. It can be observed that for effluent BF, after 60 min of electrolysis with a current density of 30 mA cm^{-2}, the cost of the energy necessary for complete COD removal was around US$ 38/kg COD; while for effluent AF, with the same current density, after 20 min of electrolysis, the energy cost was US$ 28/kg COD.

Table 4: Electrolysis time, specific energy consumption and energy cost for complete COD removal from effluents BF and AF with different current densities. Anodic area: 107 cm², flow rate: 0.54 mL s^{-1}, T: 25 °C

Parameters	i = 30 mA cm^{-2}	
	BF	AF
Time (min)	60	20
COD (mg L^{-1})	0	0
Energy consumption (kW h kg COD^{-1})	40.2	29.8
Energy cost (US$. kgCOD^{-1})	37.8	28.0

Initial concentration of COD: BF = 878.79 mg L^{-1} and AF = 712.12 mg L^{-1}.

Cañizares et al. [42] compared the efficiency and costs of three available technologies (conductive-diamond electrochemical oxidation, ozonation and Fenton) for olive mill wastewaters treatment. They showed that only conductive-diamond electrochemical oxidation was able to achieve a complete COD removal, but the cost analysis proved that the operating cost for the Fenton process was lower than both diamond electrochemical oxidation and ozonation. For comparison, the cost of a 1 kg COD removal by electrochemical oxidation was around US$ 34. On the other hand, the cost for the treatment of effluent AF (in this work) was US$ 28/kg COD for its complete removal and US$ 9.8/kg COD for disposal according to the limits established by environmental legislation of the State of Rio de Janeiro/Brazil for liquid effluent from industrial processes.

CONCLUSIONS

Results shown that Ti/RuO$_2$ electrode can be an efficient alternative for the treatment of effluents containing residues of petroleum and petroleum products, such as oils and greases. With a current density of 10 mA cm^{-2} and after 120 min of electrolysis, a COD removal above 96% for effluent AF was achieved, while for effluent BF the removal was 87%, under the same experimental conditions.

A total COD removal from effluents BF and AF was achieved when the current density was increased from 10 to 30 mA cm^{-2}. Under these conditions, after 60 min of electrolysis a complete COD removal from effluent BF was obtained, while for the effluent AF only 30 min of electrolysis were necessary.

An increase of current density led to a decrease of electrolysis time for the complete COD removal from both effluents BF and AF. However, the increase of current density caused an increase of specific energy consumption. For effluent BF, the energy cost necessary to achieve a complete COD removal with a current density of 30 mA cm^{-2} after 60 min of electrolysis was around US$ 38/kg COD, while for effluent AF, after only 30 min of electrolysis it was US$ 28/kg COD.

ACKNOWLEDGMENTS

The authors are grateful to CAPES for the financial support and CENPES/PETROBRAS/TRANSPETRO for supplying the samples.

REFERENCES

1. A. Szklo, V.C. Uller, Fundamentos do Refino de Petroleo – Economia, second ed., Interciencias, Rio de Janeiro, 2008.
2. G. Li, T. An, J. Chen, G. Sheng, J. Fu, F. Chen, S. Zhang, H. Zhao, Photoelectrocatalytic decontamination of oilfield produced water containing refractory organic pollutants in the presence of high concentration of chlorideions, J. Hazard. Mater. B138 (2006) 392–400.
3. Y. Yavuz, A.S. Koparal, U.B. Ogutveren, Treatment of petroleum refinery wastewater by electrochemical methods, Desalination 258 (2010) 201–205.

4. F. Ahmadun, A. Pendashteh, L.C. Abdullah, D.R.A. Biak, S.S. Madaeni, Z.Z. Abidin, Review of technologies for oil and gas produced water treatment, J. Hazard. Mater. 170 (2009) 530–551.
5. A.M.Z. Ramalho, C.A. Martinez-Huitle, D.R. Silva, Application of electrochemical technology for removing petroleum hydrocarbons from produced water using a DSA-type anode at different flow rates, Fuel 89 (2010) 531–534.
6. M. Santander, R.T. Rodrigues, J. Rubio, Modified jet flotation in oil (petroleum) emulsion/water separations, Colloids Surf. A Physicochem. Eng. Asp. 375 (2011) 237–244.
7. B. Tansel, B. Pascual, Removal of emulsified fuel oils from brackish and pond water by dissolved air flotation with and without polyelectrolyte use: pilotscale investigation for estuarine and near shore applications, Chemosphere 85 (2011) 1182–1186.
8. C.E. Santo, V.J.P. Vilar, C.M.S. Botelho, A. Bhatnagar, E. Kumar, R.A.R. Boaventura, Optimization of coagulation–flocculation and flotation parameters for the treatment of a petroleum refinery effluent from a Portuguese plant, Chem. Eng. J. 183 (2012) 117–123.
9. A.I. Zouboulis, A. Avranas, Treatment of oil-in-water emulsions by coagulation and dissolved-air flotation, Colloids Surf A Physicochem. Eng. Asp. 172 (2000) 153–161.
10. Y. Sun, Y. Zhang, X. Quan, Treatment of petroleum refinery wastewater by microwave-assisted catalytic wet air oxidation under low temperature and low pressure, Sep. Purif. Technol. 62 (2008) 565–570.
11. D.B. Hasan, A.R.A. Aziz1, W.M.A.W. Daud, Oxidative mineralisation of petroleum refinery effluent using Fenton-like process, Chem. Eng. Res. Des. 90 (2012) 298–307.
12. A. Coelho, A.V. Castro, M. Dezotti, G.L. Sant'Anna Jr., Treatment of petroleum refinery sourwater by advanced oxidation processes, J. Hazard. Mater. B137 (2006) 178–184.

13. D.E.C. Mazzeo, C.E. Levy, D.F. Angelis, M.A. Marin-Morales, BTEX biodegradation by bacteria from effluents of petroleum refinery, Sci. Total Environ. 408 (2010) 4334–4340.
14. R.M. Perez, G. Cabrera, J.M. Gomez, A. Abalos, D. Cantero, Combined strategy for the precipitation of heavy metals and biodegradation of petroleum in industrial wastewaters, J. Hazard. Mater. 182 (2010) 896–902.
15. Y. Feng, Y. Cui, B. Logan, Z. Liu, Perfomance of Gd-doped Ti-based Sb-SnO2 anodes for electrochemical destruction of phenol, Chemosphere 70 (2008) 1629–1639.
16. Y. Yavuz, A.S. Koparal, Electrochemical oxidation of phenol in a parallel plate reactor using ruthenium mixed metal oxide electrode, J. Hazard. Mater. B136 (2006) 296–302.
17. X. Lia, Y. Cui, Y. Feng, Z. Xie, J. Gu, Reaction pathways and mechanisms of the electrochemical degradation of phenol on different electrodes, Water Res. 39 (2005) 1972–1981.
18. O. Scialdone, S. Randazzo, A. Galia, G. Silvestri, Electrochemical oxidation of organics in water: Role of operative parameters in the absence and in the presence of NaCl, Water Res. 43 (2009) 2260–2272.
19. B.K. Korbahti, A. Tanyolac, Electrochemical treatment of simulated industrial paint wastewater in a continuous tubular reactor, Chem. Eng. J. 148 (2009) 444–451.
20. D. Ghernaout, M.W. Naceur, A. Aouabed, On the dependence of chlorine byproducts generated species formation of the electrode material and applied charge during electrochemical water treatment, Desalination 270 (2011) 9–22.
21. G. You, J. Wang, Laboratory study of the electrochemical pre-oxidation for improving thermodynamic stability of an oilfield produced water, J. Petroleum Sci. Eng. 76 (2011) 51–56.
22. X. Xing, X. Zhu, H. Li, Y. Jiang, J. Ni, Electrochemical oxidation of nitrogenheterocyclic compounds at boron-doped diamond electrode, Chemosphere 86 (2012) 368–375.

23. H. Ma, B. Wang, Electrochemical pilot-scale plant for oil field produced wastewater by M/C/Fe electrodes for injection, J. Hazard. Mater. B132 (2006) 237–243.
24. M.R.G. Santos, M.O.F. Goulart, J. Tonholo, C.L.P.S. Zanta, The application of electrochemical technology to the remediation of oily wastewater, Chemosphere 64 (2006) 393–399.
25. D. Pletcher, F.C. Walsh, Industrial Electrochemistry, second ed., Chapman and Hall, London, 1982.
26. G. Chen, Electrochemical technologies in wastewater treatment, Sep. Purif. Tchnol. 38 (2004) 11–41.
27. D.D.C. Freire, G.L. Sant'anna Jr., A proposed method modification for the determination of COD saline waters, Environ. Technol. 19 (1998) 1243–1247.
28. Y. Feng, X. Li, Electro-catalytic oxidation of phenol on several metal–oxide electrodes in aqueous solution, Water Res. 37 (2003) 2003.
29. M. Ferreira, H. Varela, R.M. Torresi, G. Tremiliosi-Filho, Electrodes passivation caused by polymerization of different phenolic compounds, Electrochim. Acta. 52 (2006) 434–442.
30. M.H. Zareie, B.K. Korbahti, A. Tanyolac, Non-passivation polymeric structures in electrochemical conversion of phenol in the presence of NaCl, J. Hazard. Mater. 87 (2001) 199–212.
31. J.D. Rodgers, W. Jedral, N.J. Bunce, Electrochemical oxidation of chlorinated phenols, Environ. Sci. Technol. 33 (1999) 1453–1457.
32. J. Iniesta, J. Gonzalez-Garcia, E. Exposito, V. Montiel, A. Aldaz, Influence of chloride ion on electrochemical degradation of phenol in alkaline medium using bismuth doped and pure PbO_2 anodes, Water Res. 35 (2001) 3291–3300.
33. E. Chatzisymeon, S. Fierro, I. Karafyllis, D. Mantzavinos, N. Kalogerakis, A. Katsaounis, Anodic oxidation of phenol on Ti/IrO_2 electrode: Experimental studies, Catalysis Today 151 (2010) 185–189.
34. M. Panizza, G. Cerisola, Electrochemical degradation of gallic acid on a BDD anode, Chemosphere 77 (2009) 1060–1064.

35. I.D. Santos, A.J.B. Dutra, J.C. Afonso, Behavior of a Ti/RuO2, anode in concentrated chloride medium for phenol and their chlorinated intermediates electrooxidation, Sep. Purif. Technol. 76 (2010) 151–157.
36. C. Berrıos, R. Arce, M.C. Rezende, M.S. Ureta-Zanartu, C. Gutierrez, Electrooxidation of chlorophenols at a glassy carbon electrode in a pH 11 buffer, Electrochim. Acta 53 (2008) 2768–2775.
37. L. Yan, H. Ma, B. Wang, Y. Wang, Y. Chen, Electrochemical treatment of petroleum refinery wastewater with three-dimensional multi-phase electrode, Desalination 276 (2011) 397–402.
38. M. Lu, Z. Zhang, W. Yu, W. Zhu, Biological treatment of oilfield-produced water: A field pilot study, Int. Biodeterior. Biodegrad. 63 (2009) 316–321.
39. F. Ahmadun, A. Pendashteh, L.C. Abdullah, D.R.A. Biak, S.S. Madaeni, Z.Z. Abidin, Review of technologies for oil and gas produced water treatment, J. Hazard. Mater. 170 (2009) 530–551.
40. M. Panizza, C.A. Martinez-Huitle, Role of electrode materials for the anodic oxidation of a real landfill leachate – Comparison between Ti–Ru–Sn ternary oxide, PbO2 and boron-doped diamond anode, Chemosphere 90 (2013) 1455–1460.
41. DZ-205.R-6 – Diretriz de controle de carga organica em efluentes liquidos de origem industrial Padroes de lancamento padroes de lancamento de efluentes liquidos de efluentes liquidos do Estado do Rio de Janeiro, DZ-205.R-6, FEEMA, Rio de Janeiro, Brazil, 2007.
42. P. Canizares, R. Paz, C. Saez, M.A. Rodrigo, Electrochemical oxidation of phenolic wastes with boron-doped diamond anodes, J. Environ. Manage. 90 (2009) 410–420.

Chapter 7

Analysis of Petroleum-contaminated Soils by Diffuse Reflectance Spectroscopy and Sequential Ultrasonic Solvent Extraction–gas Chromatography

Reuben N. Okparanma, Frederic Coulon, and Abdul M. Mouazen

Department of Environmental Science and Technology, Cranfield University, Cranfield, MK43 0AL Bedfordshire, United Kingdom

ABSTRACT

In this study, we demonstrate that partial least-squares regression analysis with full cross-validation of spectral reflectance data estimates the amount of polycyclic aromatic hydrocarbons in petroleum-contaminated tropical rainforest soils. We applied the approach to 137 field-moist intact soil samples collected from three oil spill sites in Ogoniland in the Niger Delta province (5.317°N, 6.467°E), Nigeria. We used sequential ultrasonic solvent extraction–gas chromatography as the reference chemical method. We took soil diffuse reflectance spectra with a mobile fibre-optic visible and near-infrared spectrophotometer (350–2500 nm). Independent validation of combined data from studied sites showed reasonable prediction precision (root-mean-square error of prediction = 1.16–1.95 mg kg^{-1}, ratio of prediction deviation = 1.86–3.12, and validation r^2 = 0.77–0.89). This suggests that the methodology may be useful for rapid assessment of the spatial variability of polycyclic aromatic hydrocarbons in petroleum-contaminated soils in the Niger Delta to inform risk assessment and remediation.

INTRODUCTION

Rapid measurement of polycyclic aromatic hydrocarbons (PAHs) in soils can reduce the costs associated with their management, and the likely impacts of future pollution incidents through early identification. For effective management of PAHs in the environment, knowledge of their concentration and compositional distribution in the environment is vital. The information is essential to be able to identify the origin or source of the PAH (Hites and Gschwend, 1982, Okoro and Ikolo, 2007 and Yunker and MacDonald, 1995) for risk-based assessment and remediation purposes (Askari and Pollard, 2005). In the analysis of petroleum-contaminated soils for PAH, gas chromatography–mass spectrometry (GC–MS) is mostly preferred because of their relative selectivity and sensitivity (Brassington et al., 2010 and Wang and Fingas, 1995). Nevertheless, GC–MS method is relatively expensive, involves time-consuming sample preparation protocols, and relies on the use of noxious extraction solvents (Okparanma and Mouazen, 2013a). This has prompted increased research activity recently by scientists to evolve simpler, faster, and cost-effective methods of measuring PAH in contaminated soils to complement the conventional methods.

Among these innovative methods are techniques based on vibrational spectroscopy, particularly those employing diffuse reflectance sample interfaces. These techniques include Fourier transform infrared (FTIR) spectroscopy and visible and near-infrared (VIS–NIR) spectroscopy. Diffuse reflectance in the IR spectrum came to limelight in 1976, and became known as diffuse reflectance infrared Fourier transform (DRIFT) spectroscopy (Willey, 1976). Before its advent in the IR spectrum, diffuse reflectance has been commonly available in the visible and NIR spectrum (Willey, 1976). Early applications of DRIFT have been widely reported (e.g., Fuller and Griffiths, 1978). Both DRIFT and VIS–NIR are very much similar in many ramifications, but maybe more importantly are the differences between the two of them. Both techniques are now available in portable devices that can be deployed for in-field measurements without sample preparation. In terms of capital equipment cost, the prices of their latest portable models are comparable. The commercially available DRIFT (portable) system developed by Forrester et al. (2011), the 4100 ExoScan FTIR (Agilent Technologies, CA, USA), currently costs $61 301 while the latest portable model of VIS–NIR analyser, the LabSpec 5000 (Analytical Spectral Devices, Inc., CO, USA), costs $61 078 (Okparanma and Mouazen, 2013a). The former system predicted TPH with root-mean-square error of 903 mg kg^{-1} in cross-validation for TPH ranging from 0 to 11 000 mg kg^{-1} (Forrester, 2010). On the other hand, VIS–NIR spectroscopy has been successfully used for on-line (tractor-mounted) measurement of soil properties (e.g., Mouazen et al., 2007a and Kuang and Mouazen, 2013), which opens the possibility for on-line measurement of PAHs using the technology. So far, no reports about the use of DRIFT for on-line measurement can be found in the literature.

In VIS–NIR spectroscopy (350–2500 nm), energy absorption by hydrocarbon derivatives is due to overtones and combinations of fundamental vibrational C–H stretching modes of saturated CH_2 and terminal CH_3, or aromatic C–H (ArCH) functional groups (Aske et al., 2001). The difficulty in interpreting VIS–NIR spectra because of broad and overlapping bands (Stenberg, 2010), to a large extent has been overcome by the use of advanced chemometrics and data-processing techniques (Pasquini, 2003). In the analysis of spectroscopic data, multivariate calibration generally solves the problem of interference from compounds closely related to the analyte thereby eliminating the need for selectivity (Naes et al., 2002). VIS–NIR spectroscopy has been used in conjunction with various multivariate analytical techniques to predict TPH in contaminated soils (Chakraborty et al., 2010, Chakraborthy et al., 2012, Forrester et al., 2010, Graham, 1998, Malley et al., 1999 and Schwartz et al., 2012). However, studies on the application of the methodology to predict PAH in contaminated soils are few in literature (e.g., Bray et al., 2010 and Okparanma and Mouazen, 2013b). High false positive rates were reported for PAH prediction by Bray et al. (2010), which underscore the need for further research on the application of the approach.

On the other hand, models reported by Okparanma and Mouazen (2013b) for PAH prediction appear to be local models. This limits extrapolation of the models to studied soil types. Moreover, none of these two studies considered contaminated tropical rainforest soils in their applications. Therefore, to set up a VIS–NIR-based methodology to model VIS–NIR spectra for broader environmental application, it is essential to expand the approach to cover a wider range of soil types and environmental conditions.

The objective of this study was to evaluate the performance of VIS–NIR-based methodology in the prediction of PAH in contaminated tropical rainforest soils. To do this we used sequential ultrasonic solvent extraction–gas chromatography (SUSE–GC) as the benchmark method. Sites investigated in this study are in the tropical rainforests of the oil-rich Niger Delta province in Nigeria. To the best of our knowledge this study is the first attempt to adopt VIS–NIR spectroscopy for the determination of hydrocarbons in contaminated arable lands in Nigeria.

MATERIALS AND METHODS

Brief Description of the Study Area

Available geological data for the study area show that Ogoniland is located within the Niger Delta basin with soils that are broadly classified as tropical rainforest soils, which occur in the southern part of Nigeria (SPDC, 2006). Niger Delta province covers a total land area of 70 000 km^2 (Niger Delta Environmental Survey, 1995). According to the United States Department of Agriculture (USDA) soil taxonomic order, soils in the Niger Delta belong to the Oxisols. In this study, soils were collected from within the shallow geology (top-soils) of Ogoniland, which consists of sandy clay (UNEP, 2011). Typical ranges of soil nutrients concentrations at all soil depths reported for similar ecosystems in the Niger Delta show that nitrate-nitrogen range from 0.01 to 1.96 mg kg^{-1}, phosphorous range from 0.21 to 6.92 mg kg^{-1}, sulphate range from 0.20 to 10.91 mg kg^{-1} and soil pH range from 5.2 to 6.4 (SPDC, 2006). Total organic carbon range from 3.63 to 4.11% (Tanee and Albert, 2011).

Sample Collection

In this study, we collected soil samples from three oil spill sites located at Baraboo (4.652°N, 7.249°E), Bomu 1 (4.662°N, 7.277°E), and Bomu 2 (4.662°N, 7.249°E) all in K-Dere, Ogoniland in the Niger Delta province of Nigeria (Fig. 1). By directed sampling method, we collected soil samples from the top 15-cm depth in plastic containers and preserved them in a cooler containing ice blocks until shipment. We adopted directed sampling method because of the need to cover as much of the visible hot-spots in the contaminated sites as possible. Sample management was done according to the standards of the Nigerian Government's Department of Petroleum Resources (2002). We carried out both reference SUSE–GC analysis and VIS–NIR optical scanning of soil samples in the Environmental Analytical Facility of Cranfield University, United Kingdom.

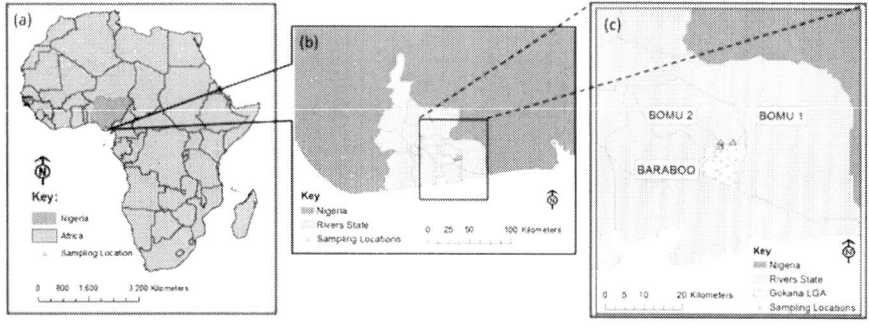

Figure 1: Sampling locations: (a) Nigeria, (b) Rivers State in the Niger Delta province, and (c) Gokana Local Government Authority in Ogoniland (Datum and Projection: GCS WGS 1984. Shape files source: ESRI®, CA, USA).

Reference Chemical Analysis of PAHs

We used SUSE–GC method to determine the concentrations of PAH compounds in the soil samples as described by Risdon et al. (2008). But, in place of acetone we used dichloromethane (Rathburn Chemicals Ltd., Walderburn, UK). Analysis of PAHs was carried with a 6890N Network GC system coupled to a 5973 Network mass selective detector (Agilent Technologies Inc., USA) operated at 70 eV in positive ion mode. Each PAH compound was quantified by the internal standard method. In this study, we adopted the same limit of quantitation (LOQ) of 0.02 mg kg^{-1} routinely used by laboratories in Nigeria for PAH quantitation in Nigerian soils because our soil samples were collected from Nigeria. LOQ is the lowest concentration at which an analyte can be reliably detected (Mitra, 2003). Consequently, concentrations less than the LOQ were removed from the total PAH computation as they were considered unreliable. A priori 5-level calibration was carried out with a calibration solution mix. The calibration solution mix was made up with EPA 525 PAH Mix-A standard solution (Sigma–Aldrich Co. Ltd., Dorset, UK), surrogate standard solution mix, and solution mix of deuterated PAHs as internal standards. Deuterated PAHs used were Naphthalene-d8, Anthracene-d10, Chrysene-d12, and Perylene-d12 while surrogates used were 2-Fluorobiphenyl and o-Terphenyl (Sigma–Aldrich Co. Ltd., Dorset, UK). Matrix spikes, duplicates, solvent and method blanks were also analysed as quality control samples.

Accuracy, Precision, and Experimental Uncertainty of Reference PAH Analytical Method

Accuracy of the reference SUSE–GC chemical method was evaluated from the percent recovery of the surrogate spiked into the sample prior to extraction according to equation (1).

$$\% \text{ Recovery} = 100 \times \frac{\text{Measured concentration}}{\text{Theoretical concentration}} \quad (1)$$

Where, theoretical concentration is equal to the concentration of the surrogate standard spiked into the sample. PAH final concentrations were bias-corrected since percent spike recoveries were between 0 and 100 as recommended in literature (Patnaik, 1997). Biases in measured concentrations of 2- and 3-ring PAHs were corrected with the percent recovery of 2-Fluorobiphenyl while the percent recovery of o-Terphenyl was used to correct for biases in measured concentrations of 4-, 5-, and 6-ring PAHs. Individual PAH concentrations were then summed up to get total PAH. Since several replicate analyses of samples were not possible in this study, reproducibility (precision) of the analytical method was determined by estimating the relative percent difference of duplicate analyses of one sample in each batch of samples as recommended in literature (Patnaik, 1997). Relative percent difference was deduced by means of equation (2).

$$\text{Relative percent difference (\%)} = \frac{(a_1 - a_2)}{\frac{(a_1 + a_2)}{2}} \times 100 \qquad (2)$$

Where a_1 and a_2 are the PAH concentrations in the first and second duplicate sample, respectively. For each batch of samples, experimental uncertainty was estimated using confidence interval at 95% level of confidence for duplicate analyses of one randomly chosen sample. The confidence interval was deduced by the well-known expression in equation (3).

$$\text{Confidence interval} = \bar{x} \pm \frac{ts}{\sqrt{n}} \qquad (3)$$

Where, $\bar{x}$ = the sample mean, t = Student's t for a desired level of confidence, s = the sample standard deviation, and n = the number of measurements. Then, percent error (absolute) in the current measurement was checked against PAH results obtained from a commercial laboratory in Nigeria. To do this, we ran duplicate analyses of the same sample from Baraboo site that was previously analysed in the commercial laboratory. Percent error was then computed by means of equation (4).

$$\% \text{ Error} = \left| \frac{T - E}{T} \right| \times 100 \qquad (4)$$

Where, T = "known" PAH value from commercial laboratory, and E = measured mean PAH value from current study.

Optical Measurement of Soil Samples

We took diffuse reflectance spectra from the soil samples with a mobile fibre-optic LabSpec2500® VIS–NIR spectrophotometer (350–2500 nm) coupled to a high-intensity probe (Analytical Spectral Devices Inc., CO, USA). The spectrophotometer has one Si array (350–1000 nm) and two Peltier-cooled InGaAs detectors (1000–1800 nm and 1800–2500 nm). Spectral sampling interval of the instrument was 1 nm across the entire spectral range. However, the spectral resolution was 3 nm at 700 nm and 10 nm at 1400 and 2100 nm. The high-intensity probe has a built-in light source made of a quartz-halogen bulb of 2727 °C. The light source and detection fibres are assembled in the high-intensity probe enclosing a 35-degree angle. Before use, and after every 30 min, the instrument was optimised by white-referencing with a white Spectralon disc of almost 100% reflectance. Reflectance spectra were taken from each soil sample, tightly packed and levelled out in a cuvette, at three different positions, 120° apart. Each sample was scanned three times at each position and averaged before spectral pre-processing and multivariate analysis.

Spectral Pre-processing

The pre-processing of soil spectra was carried out with the Unscrambler® X (CAMO Software AS, Oslo, Norway). Noisy portions at the extremes of the spectrum (i.e., from 350 to 449 nm and 2451–2500 nm) were removed due to low instrument sensitivity at these wavelengths. Spectral truncation was followed by smoothing (by averaging successive 5-nm wavelengths), leaving a total of 401 wavelengths in the working range of 452–2450 nm. Then spectral transformation was carried out by successive combination of maximum normalization and Savitzky–Golay first derivative of polynomial order of two and two smoothing points. Normalization helps to bring all data to approximately the same scale or to get a more even distribution of the variances and the average values. First derivative removes additive baseline shifts in the data and smoothing reduces the impact of noise. These measures were aimed at reducing spurious peaks that do not hold physical or chemical information (Aske et al., 2001 and Naes et al., 2002).

Partial Least-squares (PLS) Regression Analysis

Before calibration, spectral reflectance (R) was transformed to the logarithm of the relative intensity (1/R), or absorption (Naes et al., 2002). The PLS regression analysis combines both the independent variables (reference values of PAH) and the dependent variables (wavelengths) using them as regression generators for the independent variables (Maleki et al., 2007). Detailed information about the PLS can be found in Martens and Naes (1989). We used PLS regression analysis with full cross-validation to relate the variation in a single-component variable (e.g., PAH) to the variation in a multi-component variable (e.g., wavelength) by means of Unscrambler® X. The optimal number of latent factors for future predictions was determined on the basis of the number of factors with the smallest total residual validation Y-variance or highest total explained validation Y-variance (CAMO Software, 2012). Site-specific calibration models were developed with VIS–NIR spectral data and reference SUSE–GC chemical data for each site. Also, a generalised model was developed for all three sites with the entire dataset. To develop the generalised model, 78% of the samples were used for cross-validation (calibration) while the remaining 22% were used for validation (prediction). The ratio of calibration/validation samples was chosen to ensure an equal representation of samples in the validation set by randomly choosing ten samples from each site.

Statistical Evaluation of Model Performance

Criteria used to evaluate the quality of the models were based on the root-mean-square error (RMSE) of cross-validation and prediction (equations (5) and (6), respectively), ratio of prediction deviation (RPD) (equation (8)), and corresponding coefficient of determination (r^2) (Williams and Sobering, 1986). Model performance classification system adopted was based on the following criteria: very poor model predictions if RPD < 1.0; poor if 1.0 ≤ RPD < 1.4; fair if 1.4 ≤ RPD < 1.8; good if 1.8 ≤ RPD < 2.0; very good if 2.0 ≤ RPD < 2.5; and excellent if RPD > 2.5 (Viscarra Rossel et al., 2006a).

$$RMSECV = \sqrt{\frac{\sum_{i=1}^{N} \left(\hat{y}_{CV,i} - y_i\right)^2}{N}} \tag{5}$$

$$\text{RMSEP} = \sqrt{\frac{\sum_{i=1}^{N_P}(\hat{y}-y_i)^2}{N_P}} \qquad (6)$$

$$\text{RMSE} = \sqrt{\text{MSE}} = \sqrt{E(\hat{y}-y)^2} \qquad (7)$$

$$\text{RPD} = \frac{\text{SD}}{\text{RMSE}} \qquad (8)$$

Where $\hat{y}_{CV,i}$ = estimate for y_i based on the calibration equation with i deleted; $\hat{y}$ and y_i = predicted and measured reference values respectively; $E(.)$ = statistical expectation (average) over the population of future samples; N = number of samples in the set; and SD = standard deviation of the measured reference values.

Outlier Detection Techniques

During model calibration, a priori outlier detection techniques were adopted to remove influential X- and Y-outliers. The influence of sample outliers was ascertained using the influence plot (Fig. 2). This was done by plotting the residual X- and Y-variances against leverages. Samples with a high leverage and high residual X- or Y-variance were considered as potential outliers (CAMO Software, 2012). Then the projected influence plots were used to confirm those samples with high residuals. Nonetheless, before such samples were treated as outliers, they were studied in more details by marking them one-by-one and plotting the X–Y relation outliers for several model factors to observe their influence on the shape of the X–Y relationship, as recommended in literature (CAMO Software, 2012).

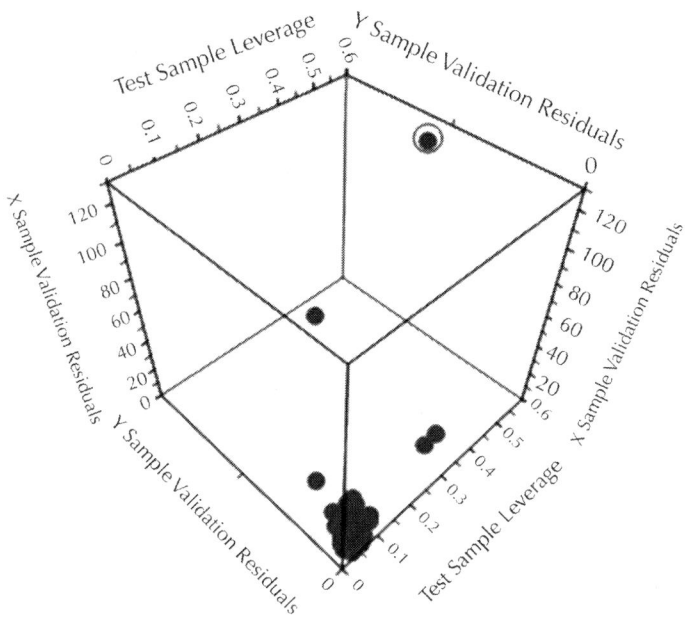

Figure 2: Detection of outliers after partial least-squares regression analysis. A potential sample outlier, marked in circle, detected among samples from Baraboo site in Ogoniland, Niger Delta province of Nigeria is shown as an example.

RESULTS AND DISCUSSION

Accuracy, Precision and Level of Uncertainty in Reference SUSE–GC Analysis of PAH

Table 1 shows the statistical results of the chemical SUSE–GC analysis showing the sum of individual PAHs quantified for each site. The high spatial variability of the compounds within and among sites is clearly demonstrated in the table. It is important that the sum of a PAH fraction in the entire sample set (Table 1) is differentiated from the sum of the entire PAH fraction in a sample (Data not shown due to the number of samples involved). The latter is the total PAH in a sample used for model development. In this study, total PAH concentration is the sum of the 13 individual PAH concentrations in a sample. The 13 PAHs in Table 1 were the relevant compounds contained in the PAH standard solution used for prior instrument calibration. The concentrations of the lower boiling 2- to 4-ring PAHs (i.e., acenaphthylene to chrysene) were higher than the higher boiling 5- to 6-ring PAHs. The difficulty in quantitation increased with boiling point. The lower boiling PAHs eluted the column before the higher boing PAHs because of their shorter retention times. Overall, Bomu 2 appeared to be the most contaminated site with the highest mean total PAH (5.39 mg kg^{-1}) compared to either Baraboo site (3.38 mg kg^{-1}) or Bomu 1 (4.47 mg kg^{-1}). These values can also be confirmed from Table 1 by adding the respective mean values.

Table 1: Statistics of the chemical analysis result showing the sum of individual polycyclic aromatic hydrocarbons (PAHs) quantified for each site by reference sequential ultrasonic solvent extraction–gas chromatography (SUSE–GC)

PAH	LOQ (mg kg^{-1})	Baraboo						Bomu 1						Bomu 2					
		N	Min. (mg kg^{-1})	Max. (mg kg^{-1})	Mean (mg kg^{-1})	Range (mg kg^{-1})		N.	Min. (mg kg^{-1})	Max. (mg kg^{-1})	Mean (mg kg^{-1})	Range (mg kg^{-1})		N	Min. (mg kg^{-1})	Max. (mg kg^{-1})	Mean (mg kg^{-1})	Range (mg kg^{-1})	
Acenaphthylene	0.02	43	0.07	3.06	0.83	2.98		58	0.02	2.40	0.46	2.38		36	<0.02	8.53	1.41	8.53	
Fluorene	0.02	43	<0.02	4.80	0.73	4.80		58	<0.02	2.70	0.45	2.70		36	<0.02	6.25	0.88	6.25	
Phenanthrene	0.02	43	<0.02	3.64	0.47	3.64		58	0.02	3.15	0.81	3.13		36	<0.02	6.86	0.76	6.86	
Anthracene	0.02	43	<0.02	4.22	0.41	4.22		58	<0.02	2.57	0.36	2.57		36	<0.02	6.86	0.75	6.86	
Pyrene	0.02	43	<0.02	1.03	0.30	1.03		58	<0.02	2.83	0.56	2.83		36	<0.02	1.73	0.52	1.73	
Benz[a]anthracene	0.02	43	<0.02	0.88	0.18	0.88		58	<0.02	2.61	0.39	2.61		36	<0.02	1.01	0.15	1.01	
Chrysene	0.02	43	<0.02	1.17	0.18	1.17		58	<0.02	1.07	0.22	1.07		36	<0.02	0.55	0.09	0.55	
Benz[b]fluoranthene	0.02	43	<0.02	0.15	0.03	0.15		58	<0.02	1.33	0.23	1.33		36	<0.02	0.18	0.04	0.18	
Benzo[k]fluoranthene	0.02	43	<0.02	0.15	0.03	0.15		58	<0.02	1.23	0.19	1.23		36	<0.02	0.18	0.04	0.18	
Benzo[a]pyrene	0.02	43	<0.02	0.73	0.17	0.73		58	<0.02	1.80	0.14	1.80		36	<0.02	9.36	0.68	9.36	
Indeno[1,2,3-cd]pyrene	0.02	43	<0.02	0.22	0.02	0.22		58	<0.02	1.48	0.23	1.48		36	<0.02	0.82	0.06	0.82	
Dibenzo[a,h]anthracene	0.02	43	<0.02	0.15	0.03	0.15		58	<0.02	1.82	0.22	1.82		36	<0.02	<0.02	–	–	
Benzo[g,h,i]perylene	0.02	43	<0.02	0.22	0.02	0.22		58	<0.02	1.48	0.22	1.48		36	<0.02	0.33	0.02	0.33	

LOQ, Limit of quantitation. It is the lowest concentration at which an analyte can be reliably detected (Mitra, 2003).

N, Number of samples.

PAH, Polycyclic aromatic hydrocarbons.

Percent recoveries of spiked surrogates for the three sites, as shown in Table 2, are within the acceptable range of 40–120% (USEPA, 1999), which is typical of a reasonably accurate PAH extraction procedure.

Table 2: Accuracy of the reference sequential ultrasonic solvent extraction–gas chromatography (SUSE–GC) method used in the chemical analysis of polycyclic aromatic hydrocarbons (PAH) in topsoils from petroleum-contaminated sites in Ogoniland, Niger Delta province of Nigeria

Sampling site	No. of samples	Spiked (mg kg^{-1})	Measured (mg kg^{-1})[a]		% Recovery of spiked surrogates
			2-Fluorobiphenyl	o-Terphenyl	
Baraboo	43	2.50	1.36	1.33	53–54
Bomu 1	58	2.50	1.94	1.63	65–78
Bomu 2	36	2.50	1.19	1.40	48–56

a. Mean concentrations of the spiked surrogates.

Table 3 shows the reproducibility and experimental uncertainty for 95% confidence interval ($n=2$) of the reference PAH measurement procedure. The relative percent difference for duplicate samples for all sites is less than 20%, suggesting that the reference SUSE–GC method was within precision standards (Mayer, 2008). The rather large confidence intervals in Table 3 are attributed to the fewer number of measurements ($n = 2$). It is known that the confidence interval reduces with number of measurements as the sample mean approaches the true population mean (Harris, 2010). Nevertheless, the estimated confidence interval suggests with 95% confidence that the true mean PAH value from the duplicate measurements would lie within the estimated range.

Table 3: Experimental uncertainty and precision of reference sequential ultrasonic solvent extraction–gas chromatography (SUSE–GC) method used for analysis of polycyclic aromatic hydrocarbon (PAH) in topsoils from petroleum-contaminated sites in Ogoniland, Niger Delta of Nigeria. Test sample was randomly chosen for each site

Sampling site	First duplicate (mg kg^{-1})	Second duplicate (mg kg^{-1})	Relative percent difference (%)	Confidence interval
				Mean ± uncertainty[a] (mg kg^{-1})
Baraboo	3.67	3.22	13	3.45 ± 2.86
Bomu 1	4.49	3.88	14	4.18 ± 3.84
Bomu 2	2.89	2.66	8	2.78 ± 1.42

a. Experimental uncertainty for 95% confidence interval for duplicate measurements.

Inter-laboratory differences in reported PAH values are shown in Table 4. For the test soil sample, Table 4 shows that mean PAH value obtained in this study is comparable to PAH result from the commercial laboratory in Nigeria. This implies that the current PAH measurement by Cranfield University's operating procedures described above is in agreement with those of the commercial laboratory. The margin of error may be attributed to differences in operating procedures used in both laboratories (Table 4). Moreover, inter-laboratory differences in reported values of an analyte in chemical analysis are a common occurrence (Risdon et al., 2008).

Table 4: Inter-laboratory differences in reported PAH values and selected rubrics in operating procedures for test soil sample from petroleum-contaminated site in Baraboo in Ogoniland, Niger Delta of Nigeria

Analyte and selected rubrics	This study	Commercial laboratory
PAH (mg kg^{-1})	3.45 ± 2.86	2.73[a]
Percent error (%)[b]	26	–
Quantitation method	Internal standard	External standard

Extraction method	Sequential ultrasonic solvent extraction	Sonication water bath (5-h sonication)
Extracting solvent(s)	Dichloromethane and hexane (1:1)	n-Pentane
Surrogate standard(s)	2-Fluorobiphenyl and o-Terphenyl	1-Chlorooctadecane

a. Uncertainty was not reported.

b. Commercial laboratory PAH result was taken as the "known" value.

The PAH Partial Least-squares Regression Models

Table 5 summarises the statistical results of the PLS models. The cross-validation models for pre-processed spectra used smaller number of latent variables than raw spectra. Similarly, cross-validation models developed by pre-processed spectra had better quality than raw reflectance models. Overall, the quality of the cross-validation models is classified as ranging from very good to excellent on the basis of RPD values. Fig. 3 shows the linear relationship between reference SUSE–GC-measured and VIS–NIR-predicted PAH data for pre-processed spectra of entire dataset. Only three sample outliers were removed from the cross-validation datasets while none was removed from the independent validation set. On the basis of RPD values, the model quality is classified as excellent prediction. The quality of the model demonstrates the possibility of using VIS–NIR spectroscopy for quantitative determinations (Viscarra Rossel et al., 2006a). Moreover, as shown in Table 5 the similarity between cross-validation and validation standard deviations indicates that the prediction models are not skewed. These results are comparable with RPD values of 1.7 and 2.5 reported by previous researchers for TPH prediction in field-moist soils by PLS regression analysis (Chakraborty et al., 2010 and Chakraborthy et al., 2012).

Table 5: Statistical results of partial least-squares (PLS) site-specific calibration and general prediction models for polycyclic aromatic hydrocarbons (PAH) in petroleum-contaminated tropical rainforest soils from Ogoniland in the Niger Delta province of Nigeria developed by visible and near-infrared (VIS–NIR) spectroscopy

Sampling site	No. of samples	Reflectance spectra					Combined pre-processing spectra[a]				
		r^2	RMSE (mg kg^{-1})	SD	RPD	LV	r^2	RMSE (mg kg^{-1})	SD	RPD	LV
Cross-validation set											
Baraboo	43	0.78	0.82	1.81	2.20	8	0.84	0.64	1.81	2.81	6
Bomu 1	58	0.76	1.41	2.95	2.09	11	0.83	1.22	2.95	2.41	3
Bomu 2	36	0.77	1.98	4.08	2.07	5	0.83	1.42	4.08	2.87	7
General	107	0.69	1.61	3.05	1.81	11	0.82	1.30	3.05	2.34	8
Validation set											
General	30	0.77	1.95	3.63	1.86	11	0.89	1.16	3.63	3.12	8

LV, Latent variable.
RMSE, Root-mean-square error.
RPD, Ratio of prediction deviation.
SD, Standard deviation.
a. Combination of maximum normalisation, and first derivative and smoothing by Savitzky–Golay.

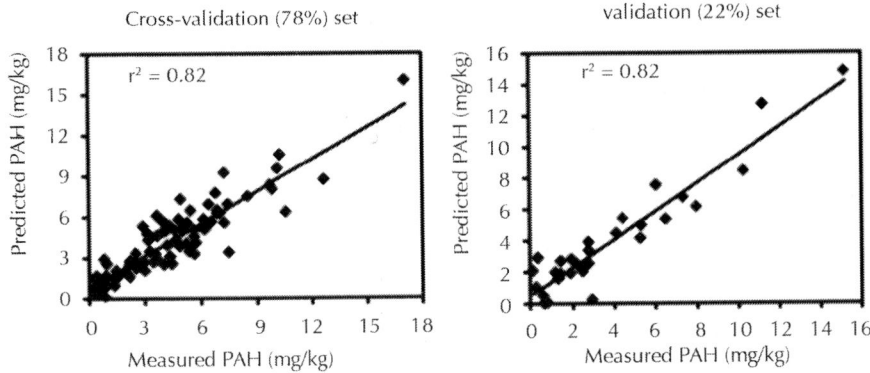

Figure 3: Scatter plots of chemically measured vs. predicted values of polycyclic aromatic hydrocarbons (PAH). These were developed by partial least-squares (PLS) regression analysis with pre-processed spectra of 137 soil samples from petroleum-contaminated tropical rainforest soils collected from Ogoniland, Niger Delta, Nigeria.

Spectral Reflectance of Petroleum-contaminated Tropical Rainforest Soils

Fig. 4 shows mean VIS–NIR spectral reflectance curves of field-moist tropical rainforest soils consisting of sandy clay from three oil spill sites in Ogoniland, Niger Delta, Nigeria. Generally, soil spectral reflectance decreased with increasing mean PAH concentration particularly in the NIR region (700–2500 nm). There was however, a slight shift in this trend for Baraboo and Bomu 1 reflectance curves in the visible range (452–700 nm). This is attributed to soil colour-associated changes in the visible range brought about when water and/or oil (e.g., diesel) is added to soil (Mouazen et al., 2005 and Okparanma and Mouazen, 2013b). In the reflectance curves for the contaminated sites, spectral absorption minima of hydrocarbon-based oil were observed around 1647, 1712, and 1752 nm in the first overtone region of the NIR band (Fig. 4). The absorption around 1647 nm is attributed to C–H stretching modes of ArCH linked to PAH. Absorptions around 1712 and 1752 nm are attributed to C–H stretching modes of terminal CH_3 and saturated

CH$_2$ groups linked to TPH, both present in the contaminating hydrocarbon-based oil (Forrester et al., 2010, Osborne et al., 1993 and Workman and Weyer, 2008). These features are practically absent in the uncontaminated reflectance curve as shown by the number of reflectance shoulders that appeared in the reflectance curves between 1452 and 1952 nm (Fig. 4).

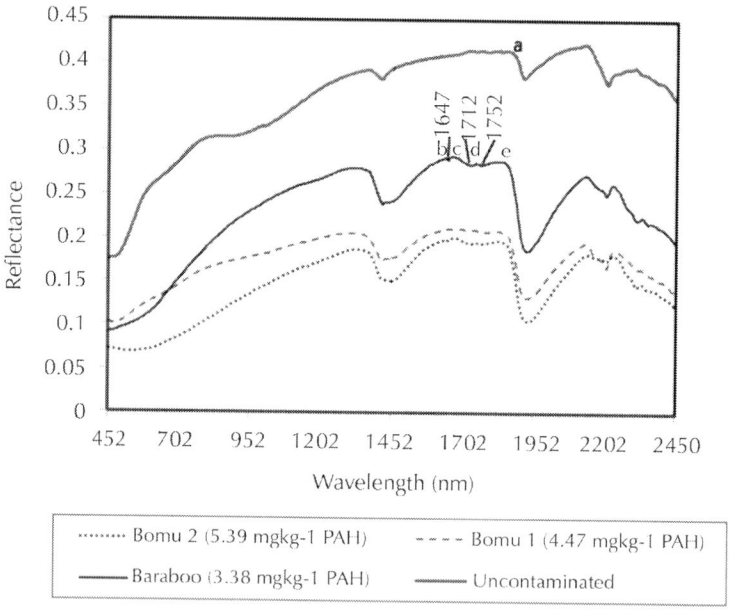

Figure 4: Mean VIS–NIR spectral reflectance curves of petroleum-contaminated tropical rainforest soils from three oil spill sites in Ogoniland in the Niger Delta province of Nigeria. Values in the legend are average PAH concentrations. Point a, b, c, d, and e are reflectance shoulders.

One reflectance shoulder appeared around 1830 nm (point a) in the uncontaminated reflectance curve while four appeared around 1630, 1675, 1737, and 1830 nm (point b, c, d, and e, respectively) in the contaminated reflectance curves (Fig. 4). A reflectance shoulder appears between two absorption wavelengths of colour, water and/or hydrocarbon (Mouazen et al., 2007b and Okparanma and Mouazen, 2013b). The reflectance shoulder at 1830 nm is attributed to water absorption in the first overtone band around 1950 nm (Stenberg, 2010) and hydrocarbon absorption in the first overtone band around 1712 nm (Osborne et al., 1993 and Workman and Weyer, 2008). The shoulder at 1630 nm is attributed to water absorption in the second overtone region around 1450 nm (Mouazen et al., 2007b) and hydrocarbon absorption in the first overtone band around 1647 nm (Workman and Weyer, 2008). The shoulder at 1675 nm is attributed to hydrocarbon absorptions in the first overtone band around 1647 and 1712 nm (Workman and Weyer, 2008). The shoulder at 1737 nm is attributed to hydrocarbon absorptions in the first overtone band around 1712 and 1752 nm (Workman and Weyer, 2008). Therefore, the hydrocarbon absorption bands around 1647 and 1712 nm differentiate the uncontaminated from contaminated reflectance curves (Fig. 4).

Regression Coefficients

Regression coefficients in PLS regression analysis provide a summary of all predictors and a given response. The plot of regression coefficients is used to identify important wavelengths for the prediction of relevant soil properties. Fig. 5 shows bar plots of regression coefficients versus wavelength derived after PLS regression analysis using raw VIS–NIR spectral data and reference SUSE–GC chemical data for each site. In the bar plots, the absolute value of the regression coefficients indicate the relative importance of the wavelength on the basis of explained X-variance in the model. In Fig. 5, negative coefficients around 1712 and 1752 nm show a positive link to absorptions due to vibrational C–H stretching modes of terminal CH_3 and saturated CH_2 functional chemical groups linked to TPH. But positive coefficients around 1647 nm are consistent with absorptions due to vibrational C–H stretching modes of ArCH functional groups linked to PAH. These results agree with absorption bands observed in the reflectance curves (see Fig. 4). Over the modelling wavelength range of 452–2450 nm, the bar plots show that intensities of regression coefficients vary in magnitude for each site. Nonetheless, the positions of important wavelengths are largely similar (Fig. 5) even though the samples are from different sites. This shows a similarity in soil property among the three sites. We observed larger regression coefficients in Fig. 5b and c than in Fig. 5a. The intensity of regression coefficients for Baraboo site was the least (Fig. 5a), and samples from Baraboo site showed the least level of contamination among the three sites as well (see Fig. 4). This may be explained by the fact that the mechanism of PAH prediction is attributed to co-variation of PAH with other soil properties that have direct spectral responses in the NIR spectral range, particularly water, clay minerals and organic carbon (Okparanma and Mouazen, 2013c). This observation was also typified in Fig. 5 by the sizes of coefficients around 950, 1450, 1950, 2200, 2212, and 2300 nm. Absorptions around 950, 1450 and 1950 nm in the NIR band are due to O–H stretching modes of water in the O–H second and first overtones, and combinations band respectively (Whalley and Stafford, 1992). Absorption features linked to metal-OH bend plus O–H stretch combinations around 2200 and 2300 nm are characteristic of clay minerals (Clark et al., 1990 and Viscarra Rossel et al., 2006b). Absorption features attributed to long-chain C–H + C–H and C–H + C–C stretch combinations around 2150 nm and 2212 nm are unique to soil organic matter (Forrester et al., 2010).

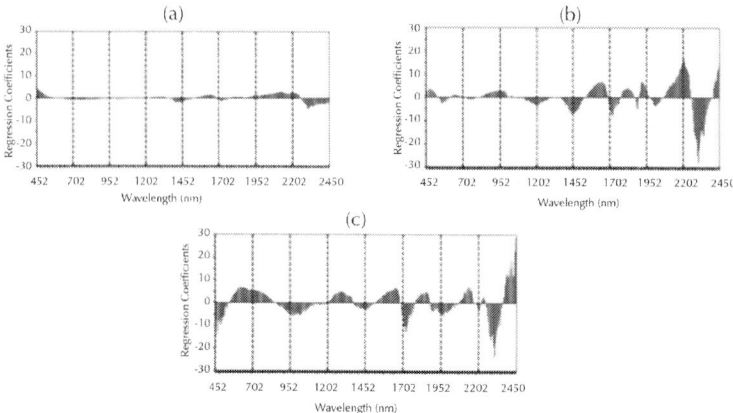

Figure 5: Plots of regression coefficients vs. wavelength after partial least squares (PLS) regression analysis by visible and near infrared (VIS–NIR)

diffuse reflectance spectroscopy for petroleum-contaminated tropical rainforest soils collected from (a) Baraboo, (b) Bomu 1, and (c) Bomu 2 in Ogoniland, Niger Delta, Nigeria.

CONCLUSIONS

In this study, results show that soil diffuse reflectance decreased with increasing PAH concentration. Positive regression coefficients around 1647 nm show a link to PAH. Additionally, the quality of generalized PLS models for PAH predictions ranged from good to excellent (validation r^2 = 0.77–0.89, RMSEP = 1.16–1.95 mg kg^{-1}, and RPD = 1.86–3.12). This demonstrates the possibility of using VIS–NIR spectroscopy and PLS regression analysis for rapid quantitative determination of PAH in petroleum-contaminated tropical rainforest soils in the Niger Delta province of Nigeria. However, it should be pointed out that the extrapolation of the model is limited to the three sites investigated in this study namely Baraboo, Bomu 1, and Bomu 2 in Ogoniland, Niger Delta region of Nigeria. To set up a model based on VIS–NIR spectroscopy for general application in the Niger Delta province of Nigeria, we recommend the use of larger dataset covering both the concentration range and all the other sources of variability in oil spill sites in the Niger Delta region.

ACKNOWLEDGMENTS

This research has been funded by the Petroleum Technology Development Fund (PTDF), Nigeria, through financial assistance in the form of doctoral studentship. The Rivers State University of Science and Technology, Port Harcourt, Nigeria, also provided support through its Academic Staff Development Program.

REFERENCES

1. Askari, K., Pollard, S.J.T., 2005. The UK Approach for Evaluating Human Health Risks from Petroleum Hydrocarbons in Soils. Science Report P5-080/TR3. Environment Agency, Rio House, Waterside Drive, Aztec West, Almondsbury, Bristol, UK, pp. 1e22.
2. Aske, N., Kallevik, H., Sjoblom, J., 2001. Determination of saturate, aromatic, resin, and asphaltenic (SARA) components in crude oils by means of infrared and near-infrared spectroscopy. Energy Fuels 15, 1304e1312.
3. Brassington, K.J., Pollard, S.J.T., Coulon, F., 2010. Weathered hydrocarbon wastes: a risk management primer. In: Timmis, K.N. (Ed.), Handbook of Hydrocarbon and Lipid Microbiology. Springer-Verlag, Berlin Heidelberg, pp. 2488e2499.
4. Bray, J.G., Viscarra Rossel, R.A., McBratney, A.B., 2010. Diagnostic screening of urban soil contaminants using diffuse reflectance spectroscopy. In: Viscarra Rossel, R.A., McBratney, A.B., Minasny, B. (Eds.), Proximal Soil Sensing. Springer-Verlag, Berlin Heidelberg, pp. 191e199.
5. CAMO Software, 2012. The Unscrambler X Version 10.2 User's Guide. CAMO Software AS, Nedre Vollgate, Oslo, Norway.
6. Chakraborthy, S., Weindorf, D.C., Zhu, Y., Li, B., Morgan, C.L.S., Ge, Y., Gulbraith, J., 2012. Spectral reflectance variability from soil physicochemical properties in oil contaminated soils. Geoderma 177e178, 80e89.
7. Chakraborty, S., Weindorf, D.C., Morgan, C.L.S., Ge, Y., Galbraith, J.M., Li, B., Kahlon, C.S., 2010. Rapid identification of oil-contaminated soils using visible near-infrared diffuse reflectance spectroscopy. J. Environ. Qual. 39, 1378e1387.
8. Clark, R.N., King, T.V.V., Klejwa, M., Swayze, G., Vergo, N., 1990. High spectral resolution reflectance spectroscopy of minerals. J. Geophys. Res. 95, 12653e12680.

9. Department of Petroleum Resources, 2002. Environmental Guidelines and Standards for the Petroleum Industry in Nigeria (EGASPIN). Ministry of Petroleum and Natural Resources, Abuja, Nigeria, p. 314.
10. Forrester, S., Janik, L., McLaughlin, M., 2010. An infrared spectroscopic test for total petroleum hydrocarbon (TPH) contamination in soils. In: Proceedings of the 19th World Congress of Soil Science, Soil Solutions for a Changing World, Brisbane, Australia, August 1e6, pp. 13e16.
11. Forrester, S., 2010. In-situ determination of total petroleum hydrocarbon (TPH) contamination: a quick infrared spectroscopic test for TPH at contaminated sites. In: Poster Presentation at the 19th World Congress of Soil Science, Soil Solutions for a Changing World, Brisbane, Australia, August 1e6.
12. Forrester, S.F., Janik, L.J., McLaughlin, M.J., 2011. Method of Contaminant Prediction. PCT Patent WO/2011/035391. Date issued: 31 March.
13. Fuller, M.P., Griffiths, P.R., 1978. Diffuse reflectance measurements by infrared Fourier transform spectrometry. Anal. Chem. 50 (13), 1906e1910.
14. Graham, K.N., 1998. Evaluation of Analytical Methodologies for Diesel Fuel Contaminants in Soil (M.Sc. thesis). The University of Manitoba, Canada (Unpublished results).
15. Harris, D.C., 2010. Quantitative Chemical Analysis, eighth ed. W.H. Freeman and Company, New York.
16. Hites, R.A., Gschwend, P.M., 1982. The ultimate fate of polycyclic aromatic hydrocarbons in marine and lacustrine sediments. In: Cooke, M., Dennis, A.J., Fisher, F.L. (Eds.), Poly-nuclear Aromatic Hydrocarbons e Physical and Biological Chemistry. Bettelle Press, Columbus, Ohio, USA, pp. 357e365.
17. Kuang, B., Mouazen, A.M., 2013. Effect of spiking strategy and ratio on calibration of on-line visible and near infrared

soil sensor for measurement in European farms. Soil Till. Res. 128, 125e136.
18. Maleki, M.R., Mouazen, A.M., Ramon, H., De Baerdemaeker, J., 2007. Optimization of soil VISeNIR sensor-based variable rate application system of soil phosphorus. Soil Till. Res. 94, 239e250.
19. Malley, D.F., Graham, K.N., Webster, G.R.B., 1999. Analysis of diesel contaminated soils by near-infrared reflectance spectroscopy and solid phase micro extraction-gas chromatography. J. Soil Contam. 8, 481e489.
20. Mayer, A.S., 2008. Lecture Note, GE3850: Geohydrology. Department of Civil and Environmental Engineering, Michigan Technology University, MI. Unpublished data.
21. Martens, H., Naes, T., 1989. Multivariate Calibration, second ed. John Wiley and Sons, Chichester, UK.
22. Mitra, S., 2003. Sample Preparation Techniques in Analytical Chemistry. Wiley and Sons, Inc., Publication, Hoboken, NJ, USA.
23. Mouazen, A.M., De Baerdemaeker, J., Ramon, H., 2005. Towards development of online soil moisture content sensor using a fibre-type NIR spectrophotometer. Soil Till. Res. 80, 171e183.
24. Mouazen, A.M., Maleki, M.R., De Baerdemaeker, J., Ramon, H., 2007a. On-line measurement of some selected soil properties using a VISeNIR sensor. Soil Till. Res. 93 (1), 13e27.
25. Mouazen, A.M., Karoui, R., Deckers, J., De Baerdemaeker, J., Ramon, H., 2007b. Potential of visible and near-infrared spectroscopy to derive colour groups utilizing the Munsell soil colour charts. Biosyst. Eng. 97, 131e143.
26. Naes, T., Isaksson, T., Fearn, T., Davies, T., 2002. A User Friendly Guide to Multivariate Calibration and Classification. NIR Publications, Chichester, UK.
27. Niger Delta Environmental Survey, 1995. Background and Mission: Briefing Note 1. Publication of the Steering Committee, NDES, Falomo, Lagos, Nigeria, pp. 1e7.

28. Okoro, D., Ikolo, A.O., 2007. Sources and compositional distribution of polycyclic aromatic hydrocarbons in soils of Western Niger Delta. J. Appl. Sci. Technol. 12 (1e2), 35e40.
29. Okparanma, R.N.,Mouazen,A.M., 2013a. Determinationof total petroleumhydrocarbon (TPH) and polycyclic aromatic hydrocarbon(PAH) in soils: a reviewof spectroscopic and non-spectroscopic techniques. Appl. Spectrosc. Rev. 48 (6), 458e486.
30. Okparanma, R.N., Mouazen, A.M., 2013b. Combined effects of oil concentration, clay and moisture contents on diffuse reflectance spectra of diesel-contaminated soils. Water Air Soil Pollut. 224 (5), 1539e1556.
31. Okparanma, R.N., Mouazen, A.M., 2013c. Visible and near-infrared spectroscopy analysis of a polycyclic aromatic hydrocarbon in soils. Scientific World Journal (Article in review).
32. Osborne, B.G., Fearn, T., Hindle, P.H., 1993. Practical NIR Spectroscopy e with Applications in Food and Beverage Analysis, second ed. Longman Group UK Limited, England.
33. Pasquini, C., 2003. Near infrared spectroscopy: fundamentals, practical aspects and analytical applications. J. Braz. Chem. Soc. 14 (2), 198e219.
34. Patnaik, P., 1997. Handbook of Environmental Analysis: Chemical Pollutants in Air, Water, Soil and Solid Wastes. CRC Press Inc., Boca Raton, FL.
35. Risdon, G.C., Pollard, S.J.T., Brassington, K.J., McEwan, J.N., Paton, G.I., Semple, K.T., Coulon, F., 2008. Development of an analytical procedure for weathered hydrocarbon contaminated soils within a UK risk-based framework. Anal. Chem. 80 (18), 7090e7096.
36. Schwartz, G., Ben-Dor, E., Eshel, G., 2012. Quantitative analysis of total petroleum hydrocarbons in soils: comparison between reflectance spectroscopy and solvent extraction by 3 certified laboratories. Appl. Environ. Soil Sci. 2012, 1e11.

37. SPDC, 2006. Environmental Impact Assessment (EIA) of Rumuekpe (OML 22) and Etelebou (OML 28) Area 3 Dimensional Seismic Survey. Publication of the Shell Petroleum Development Company (SPDC) of Nigeria Limited, Abuja, Nigeria.
38. Available: http://www.shell.com.ng/environment-society/environment-impactassessments.html (accessed 25.02.13.).
39. Stenberg, B., 2010. Effects of soil sample pretreatments and standardized rewetting as interacted with sand classes on VISeNIR predictions of clay and soil organic carbon. Geoderma 158, 15e22.
40. Tanee, F.B.G., Albert, E., 2011. Post-remediation assessment of crude oil polluted site at Kegbara-Dere Community, Gokana L.G.A. of Rivers State, Nigeria. Bioremediation Biodegradation 2. http://dx.doi.org/10.4172/2155-6199.1000122 (accessed 26.02.13.).
41. UNEP, 2011. Environmental Assessment of Ogoniland. United Nations Environment Programme (UNEP), Nairobi, Kenya. Available: http://www.unep.org (accessed 29.10.12.).
42. USEPA, 1999. Method 8270: Semi Volatile Organic Compounds (SVOCS) by Gas ChromatographyeMass Spectrometry (GCeMS). SW-846 Manual. United States Environmental Protection Agency (USEPA), Government Printing Office, Washington, DC, USA.
43. Viscarra Rossel, R.A., Walvoort, D.J.J., McBratney, A.B., Janik, L.J., Skjemstad, J.O., 2006a. Visible, near infrared, mid infrared or combined diffuse reflectance spectroscopy for simultaneous assessment of various soil properties. Geoderma 131, 59e75.
44. Viscarra Rossel, R.A., McGlynn, R.N., McBratney, A.B., 2006b. Determining the composition of mineral-organic mixes using UVeviseNIR diffuse reflectance spectroscopy. Geoderma 137, 70e82.
45. Wang, Z., Fingas, M., 1995. Differentiation of the sources of spilled oil and monitoring of the oil weathering process using

gas chromatographyemass spectroscopy. J. Chromatogr. A 712, 321e343.

46. Whalley, W.R., Stafford, J.V., 1992. Real-time sensing of soil water content from mobile machinery: options for sensor design. Comput. Electron. Agr. 7, 269e358.
47. Willey, R.R., 1976. Fourier transform infrared spectrophotometer for transmittance and diffuse reflectance measurements. Appl. Spectrosc. 30 (6), 593e601.
48. Williams, P.C., Sobering, D.C., 1986. Attempts at standardization of hardness testing of wheat. II. The near-infrared reflectance method. Cereal Foods World 31, 417e420.
49. Workman Jr., J., Weyer, L., 2008. Practical Guide to Interpretive Near-infrared Spectroscopy. CRC Press, Taylor and Francis Group, Boca Raton, FL, USA.
50. Yunker, M.B., MacDonald, R.W., 1995. Composition and origins of polycyclic aromatic hydrocarbons in the Mackenzie River and on the Beaufort Sea Shelf. J. Arct. Inst. N. Am. 48 (2), 118e129.

Chapter 8

Molecular Reconstruction of Heavy Petroleum Residue Fractions

J.J. Verstraete, Ph. Schnongs, H. Dulot, and D. Hudebine

IFP-Lyon, 69390 Vernaison, France

ABSTRACT

Molecular reconstruction techniques are methods that allow to create mixtures of molecules from partial analytical data. In this article, a two-step reconstruction algorithm will be presented. The first step, called "stochastic reconstruction" step, assumes that oil mixtures can be described by distributions of structural blocks. The choice of the blocks and distributions is based on expert knowledge. The transformation from a set of distributions into a mixture of molecules is obtained by Monte–Carlo sampling, while a genetic algorithm adjusts the parametric distributions. The second step, termed "reconstruction by entropy maximization", improves the representativeness of the set of constructed molecules by adjusting their molar fractions. The estimation of these molar fractions is carried out by maximizing an information entropy criterion under linear constraints. The two-step reconstruction algorithm allows to rebuild mixtures that resemble the petroleum fractions more closely than the approaches used previously. To illustrate the approach, the technique is applied to petroleum vacuum residue fractions.

INTRODUCTION

In petroleum refining, both the accurate design and the optimisation of conversion processes require the development of reliable kinetic models. In order to account for the differences in reactivity of the various species, more rigorous models containing molecule-based reaction pathways are needed. Such models expect a molecular description of the feed, however. For petroleum cuts boiling above the naphtha range (atmospheric gas oil, vacuum gas oil,), a molecular description can no longer be directly obtained, even though cutting-edge analytical techniques. To circumvent this lack of molecular information, such a detailed description needs to be "reconstructed" from more "global" analyses. To this aim, a series of generic algorithms were developed in order to generate a representative mixture of molecules from standardized petroleum analyses. The idea of these molecular reconstruction algorithms is to create a discrete set of molecules that mimics the properties of the petroleum cut to be represented. The application of such an algorithm to generate a representation of heavy petroleum cuts will be described and illustrated for residue and asphaltene fractions.

During previous work (Hudebine et al., 2002; Hudebine, 2003; Hudebine and Verstraete, 2004; Verstraete et al., 2004; Van Geem et al., 2007), two different algorithms were developed to generate a complex mixture of molecules from standardized petroleum analyses: a stochastic reconstruction technique and a reconstruction by information entropy maximization. Moreover, both approaches can be advantageously combined to avoid some of the drawbacks of each technique (Hudebine and Verstraete, 2004). In the present work, these methods have been extended to be able to treat heavy petroleum fractions that contain large amounts of impurities and heteroatoms.

VACUUM RESIDUE CHARACTERISTICS

The vacuum residue of a crude oil is the heaviest product that can be obtained as a result of atmospheric distillation, followed by subsequent vacuum distillation. This petroleum fraction concentrates all the high-boiling and the most complex molecules, as well as a large part of the impurities present in the crude. These impurities include sulphur (up to 6 wt% in the residue), nitrogen (0.1–2 wt. %), oxygen (0.005–1.5 wt. %), as well as traces of metals, such as nickel and vanadium.

In addition to this atomic complexity, a very large variety of chemical structures are present. The highest molecular weight components of the vacuum residues are characterised by one or several nuclei, or cores, composed of aromatic and/or naphthenic cycles, onto which aliphatic chains are grafted. In the particular case of nuclei containing both naphthenic and benzenic cycles, the naphthenic cycles are usually peripheral. Concerning the sulphur compounds Zhao et al. (2005) and Miyabashi et al. (1995) have shown that sulphide structures and thiophenic structures (thiophenes, benzothiophenes, dibenzothiophenes, naphtheno-benzothiophenes,) predominate in vacuum residues. The structures present in nitrogen compounds fall into two large classes: neutral nitrogen compounds (pyrroles, indoles, carbazoles,) and basic nitrogen compounds (amines, pyridines, quinolines, acridines,) (Revellin et al., 2005; Revellin, 2006). Although in a smaller concentration in crude oils, the majority of the oxygen compounds are found in the vacuum residue. The different oxygen-containing structures can be subdivided in two classes: neutral oxygen compounds (furanes, aldehydes, ketones, esters, alcohols,) and acid oxygen compounds (phenols, aliphatic carboxylic acids, aromatic carboxylic acids,). Even though there is a vast body of work on characterization techniques and experimental information on molecular structures of heavy petroleum fractions, the current analytical techniques do not allow to characterise the individual molecules or isomers groups of vacuum residues. Characterization techniques for heavy petroleum fractions have been well described by Altgelt and Boduszynski (1994). Many analysis methods, such as elemental analysis (C, H, S, N, O, Ni, V), average molecular weight, nuclear magnetic resonance (NMR), and infrared (IR) spectroscopy, yield only average bulk properties. Another way of characterizing residues fractions is carried out by separation techniques, such as (partial) simulated distillation, high performance liquid chromatography (HPLC), size exclusion chromatography (SEC) and solvent solubility techniques, which separate a saturates, aromatics, resins and asphaltenes fraction through a so-called SARA analysis. The asphaltenes fraction is obtained through precipitation after mixing the sample with a paraffinic solvent, such as n-pentane or n-heptane (Speight, 1999). The exact structure of asphaltenes, however, is still subject to much controversy (Sheu and Mullins, 1996; Fenistein et al., 1998; Mullins and Sheu, 1999; Groenzin and Mullins, 2000; Sheu, 2002; Murgich, 2003; Mullins et al., 2006), but they are usually represented in the form of leaflets consisting of substituted polycondensed aromatic structures.

MOLECULAR RECONSTRUCTION TECHNIQUES

As discussed before, due to their complexity, the exact molecular structures present in heavy petroleum fractions can currently not be unveiled, even by using and combining cutting-edge analytical techniques. Hence, a different methodology must be followed. Molecular reconstruction techniques allow creating mixtures of molecules that correctly represent the characteristics of real mixtures from partial analytical data. In the literature, several approaches have been described.

Many authors have chosen to represent petroleum fractions (and especially asphaltenes) by an average model molecule in 2D (Hirsch and Altgelt, 1970; Speight, 1999; Takegami et al., 1980; Suzuki et al., 1982; Ali et al., 1990 and Ali et al., 2006; Sato, 1997; Artok et al., 1999; Gauthier et al., 2008) or 3D (Faulon et al., 1990). These average model molecules are generally constructed to be consistent with the information on the various chemical functions in the mixture, provided via e.g. infrared and NMR spectrometry. This approach suggests an average description of the mixture and allows to acquire some understanding on the average structure. However, the main drawback lies in the fact that the use of average structures is inadequate for process modelling, since the polydispersity of complex mixtures, and hence the span of properties and reactivities, is lost by this representation.

Ensembles of compounds have also been used to represent petroleum fractions. Allen and Liguras (1991) selected a set of predefined molecules and modified their molar fractions in order to obtain a mixture whose properties are close to the desired analytical data. The analyses used by these authors are gas chromatography, ^{13}C NMR and ^{1}H NMR, and allowed to setup 190 linear equality constraints as a function of the molar fractions.

Khorasheh et al. (1998) start from structural group analysis that comprises elemental analysis, ^{1}H NMR and ^{13}C NMR and that allows to determine the concentrations of a selected number of structural groups by minimizing an objective function subject to linear constraints.

Quann and Jaffe, 1992 and Quann and Jaffe, 1996 suggested a similar method but the molecules are replaced by vectors of structural blocks, called "structure oriented lumping" (SOL) vectors. More recently, Jaffe et al. (2005) extended to SOL method to vacuum residua, and more specifically to the multi-core species in these heavy fractions.

Finally Neurock (1992) and Neurock et al. (1994) developed a method called "stochastic reconstruction" (SR). The main hypothesis of this approach lies in the description of petroleum cuts by means of a set of distributions of molecular structural attributes. These distributions are then sampled by a Monte–Carlo method in order to generate a large number of molecules. When coupled to an optimisation loop on the parameters of the distributions of molecular attributes, the method has been proven able to create mixtures that closely mimic the original properties of heavy asphaltene feedstocks (Trauth, 1993; Trauth et al., 1994).

During a previous work (Hudebine et al., 2002), two algorithms for molecular reconstruction have been developed in order to rebuild light cycle oil (LCO) gas oils from overall petroleum analyses. The first method was based on stochastic reconstruction (SR) through sampling of statistical distributions, the other was a novel approach, termed reconstruction by "entropy maximization" (EM). During validation, both methods have been applied to LCO gas oils because a larger number of analytical data was available for this type of feeds and the properties of the resulting mixture could thus be compared with the original data. The results have shown a reasonably good agreement, but they also illustrated some of the drawbacks of each method. Subsequent work (Hudebine and Verstraete, 2004) advantageously coupled the SR and EM methods in a two-step molecular reconstruction algorithm and validated it on LCO gas oils.

Although the two-step molecular reconstruction algorithm has already been described elsewhere (Hudebine and Verstraete, 2004), its extension to residue fractions is not an easy task. Indeed, besides carbon, hydrogen and sulphur, oxygen- and nitrogen-compounds structures also have to be represented. In the next section, a short description of this reconstruction algorithm will be presented, while insisting upon the novel features that had to be included.

EXTENSION OF THE TWO-STEP MOLECULAR RECONSTRUCTION ALGORITHM TO RESIDUE FRACTIONS

In the first step of this molecular reconstruction algorithm, a large set of representative molecules, typically several thousands of species, is created via a SR method. After creating this ensemble of molecules, the second step of the algorithm aims at improving its composition by adjusting the mole fractions of the various species via the EM method.

The stochastic reconstruction (SR) method uses a random sampling process of parametric distributions of structural blocks (polycyclic cores, rings, chains, etc.) in order to reconstruct molecules and generate an equimolar mixture of N molecules that are representative of a selected petroleum fraction. Upon creating such a mixture, its mixture properties can be calculated and compared to the properties of the petroleum cut that needs to be represented. In an optimisation loop, the parameters of the distribution functions are modified by means of a genetic algorithm in order to minimize the difference between the calculated properties and the experimental data. Fig. 1 shows a block diagram of the basic principles of the SR step. Further details on the SR method can be found elsewhere (Hudebine et al., 2002; Hudebine, 2003; Hudebine and Verstraete, 2004).

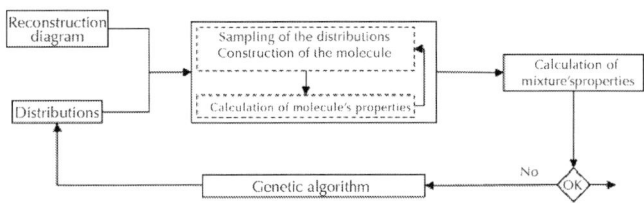

Figure 1: Block diagram of the stochastic reconstruction step.

To apply the SR step to residue fractions, a set of distributions of structural attributes needs to be chosen and a reconstruction scheme, which describes the sampling order of the different distributions, needs to be specified. The choice of the structural attributes and their distributions has been guided by knowledge of the chemical nature of these heavy products (Takegami et al., 1980; Suzuki et al., 1982; Murgich et al, 1987; Ali et al., 1990 and Ali et al., 2006; Trauth, 1993; Altgelt and Boduszynski, 1994; Miyabayashi et al., 1995; Artok et al., 1999; Mullins and Sheu, 1999; Speight, 1999; Groenzin and Mullins, 2000), but care has been taken to minimize the number of attributes.

For the residue fractions, a first distribution defines the type of molecule to be constructed (paraffins, naphthenes, aromatics/resins, or asphaltenes). For the paraffins, their chain length is determined by a gamma distribution. For non-paraffins, the number of polycyclic cores per molecule was limited to one for naphthenes, aromatics and resins, while a gamma distribution was used for the number of polycyclic cores per asphaltene molecule. For each polycyclic core, the total number of cycles per core is also characterised by a gamma distribution. Possible cycles that are considered in the SR step are benzenes, thiophenes, pyridines, pyrroles, furanes and cyclohexanes. The length of the side-chains is defined by an exponential distribution, while histograms are used to define the distributions for the presence of sulphur, nitrogen and oxygen atoms in these side-chains. The full list of distributions that are used to build a residue fraction is given in Table 1. The molecules are built by means of a reconstruction scheme that defines the sequence in which the distributions are sampled. This reconstruction scheme is schematically depicted in Fig. 2.

Table 1: Description of the structural attributes used during the SR step

	Structural attribute	Distribution	Values	Param.
1.	Type of the molecule (Par., Nap., Aro. /Res., Asp.)	Histogram	0, 1, 2 or 3	3

2.	Number of polycyclic cores in an asphaltene	Gamma	>1	2
2.	Number of cycles per core	Gamma	>1	2
4.	Number of benzenes	Gamma	>0	2
5.	Type of hetero-atomic cycles (thioph., pyr., pyrrole, furane)	Histogram	0, 1, 2 or 3	3
6.	Number of thiophenes	Histogram	1 or 2	1
7.	Number of pyridines	Histogram	1 or 2	1
8.	Number of pyrroles	Histogram	1 or 2	1
9.	Number of furanes	Histogram	1 or 2	1
10.	Alkyl chain presence on peripheral carbon atoms	Histogram	0 or 1	1
11.	Length of the paraffins	Gamma	>0	2
12.	Length of an alkyl chain (lateral/inter-core)	Exponential	>0	1
13.	A/R: sulphur substitution for aliphatic CH_2	Histogram	0 or 1	1
14.	A/R: other atom substitution for aliphatic CH_3/CH_2	Histogram	0 or 1	1
15.	Asp: sulphur substitution for aliphatic CH_2	Histogram	0 or 1	1
16.	Asp: other atom substitution for aliphatic CH_3/CH_2	Histogram	0 or 1	1
17.	Type of atom substitution (oxygen/nitrogen)	Histogram	0 or 1	1
18.	Oxygen structure (alcohol, ketone, carboxylic acid)	Histogram	0, 1 or 2	2
19.	Hydroxyl substitution on an aromatic H	Histogram	0 or 1	1

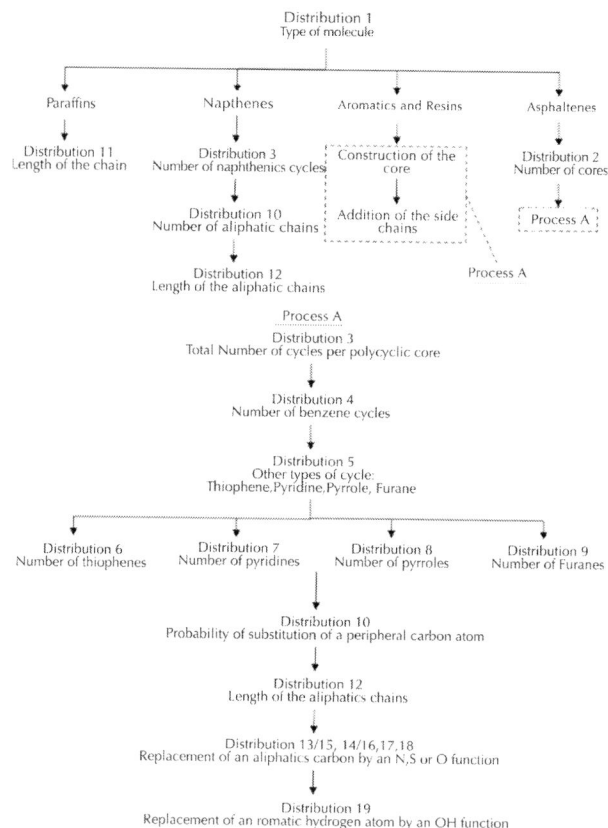

Figure 2: Reconstruction scheme for residue fractions, with process A being the construction of the polycyclic cores with its side-chains.

During the SR step, the above-described operations are applied N times so as to obtain a mixture of N molecules that is statistically representative of the distributions. For each molecule, the elemental composition, molecular weight, mass spectrum, ^{1}H NMR spectrum and ^{13}C NMR spectrum are easily obtained. For the specific gravity and the normal boiling point, group contribution methods are used (Hudebine, 2003). The properties of the equimolar mixture are subsequently calculated by linear mixing rules (elemental analysis, average molecular weight, average specific gravity, simulated distillation, mass spectrometry, ^{1}H and ^{13}C NMR analyses). In the case of the average specific gravity and the distillation, this is equivalent to the assumption of an ideal mixture behaviour. Finally, the 28 parameters of the distribution are modified in order to adjust the equimolar mixture until the desired mixture properties are found. The difference between experimental and calculated data is mathematically described by an objective function, or fitness function, which is minimized by a genetic algorithm.

Upon convergence, the equimolar set of N molecules is then passed on to the entropy maximization (EM) step, which will modify their molar fractions in order to obtain a mixture whose properties are more closely aligned to those of the petroleum cut. To adjust the molar fractions of such a large number of molecules, the optimisation problem is transformed and an information entropy criterion is maximized, which guarantees that a given molecule cannot be preferred to others when no information is available. A schematic representation of entropy maximisation is given in Fig. 3. For the underlying equations of the EM method and the mathematical solution of the optimisation problem, the reader is referred to the literature (Hudebine et al., 2002; Hudebine, 2003; Hudebine and Verstraete, 2004; Van Geem et al., 2007).

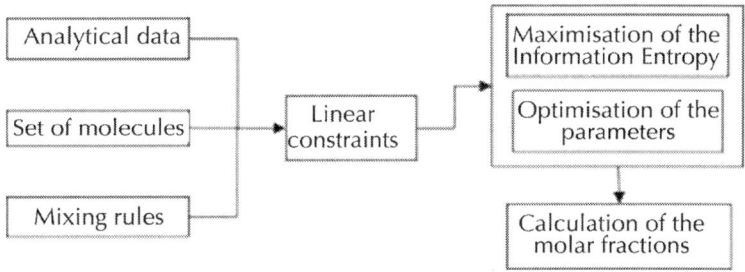

Figure 3: Block diagram of the entropy maximisation step.

RESULTS AND DISCUSSION

The two-step molecular reconstruction algorithm was applied to various atmospheric and vacuum residues, and will be illustrated in this paper for an Arabian Light vacuum residue. Table 2 lists the experimental data that was available for the vacuum residue fraction. For the IFP data, the elemental analysis in terms of C–H–N was obtained by means of thermal conductivity measurements, while O and S were obtained from infrared measurements. The nickel and vanadium contents were determined through wavelength dispersive X-ray fluorescence. For the density, ASTM D4052 was used, while the simulated distillation was obtained by means of an ASTM D5307 derived gas chromatographic method. The ^{1}H NMR analyses were carried out according to the methods described by Brown et al.

(1960) and Dickinson (1980). Concerning the SARA separation, the de-asphalting step is based on sample flocculation by contacting it with n-heptane as paraffinic solvent at 80 °C with an oil/heptane ratio of 1/50. In a second step, the SAR fractionation is conducted by a liquid chromatography method described elsewhere (Neurock, 1992; Trauth, 1993; Adam et al., 2007). The data for the Arabian Light vacuum residue were also compared to literature data taken from Neurock (1992).

Both vacuum residues have approximately the same cutpoint, and a very good agreement between the analyses was observed. The latter work also allowed to complete our analyses with an average molecular weight measurement determined by vapour pressure osmometry (Neurock, 1992; Trauth, 1993). Further details on the analytical methods can be found in Neurock (1992) and Trauth (1993). The latter two authors also separated the SARA fractions on a preparative scale, and performed an elemental analysis, molecular weight measurement and a ^{1}H NMR analysis on each of these fractions (Table 3, Table 4, Table 5 and Table 6). Combining these analytical results with the SARA yields allows to back-calculate the composition of the Arabian Light vacuum residue. As shown in Table 2, a very good agreement is found.

Table 2: Reconstruction of the Arabian Light vacuum residue: experimental data, results after the stochastic reconstruction step and the entropy maximisation step

Arabian Light vacuum residue	Experimental data (IFP)	Experimental (Neurock, 1992)	Experimental (from fractions)	Simulated (SR output)	Simulated (EM output)
Elementary An.					
Carbon (wt %)	84.7	84.7	83.7	84.4	84.7
Hydrogen (wt %)	10.2	10.0	10.8	9.9	10.2
Sulphur (wt %)	4.1	4.0	4.0	4.2	4.0
Oxygen (wt %)	0.7	1.0	1.1	1.2	0.8
Nitrogen (wt %)	0.29	0.30	0.31	0.27	0.29
Vanadium (wtppm)	120	–	–	0	0
Nickel (wtppm)	25	–	–	0	0
H/C atomic ratio (dimensionless)	1.45	1.42	1.55	1.40	1.45
S/C atomic ratio (dimensionless)	0.018	0.018	0.018	0.019	0.018
O/C atomic ratio (dimensionless)	0.006	0.009	0.010	0.011	0.007
N/C atomic ratio (dimensionless)	0.003	0.003	0.003	0.003	0.003
Density (g/ml)	1.021	–	–	–	–
Molecular weight (g/mol)	–	998	996	1017	997

¹H NMR Analysis					
Aromatic H (at %)	7.3	7.0	7.5	4.3	7.4
Alpha H (at %)	11.8	11.0	11.6	12.6	11.8
Beta H (at %)	59.6	60.8	59.7	63.1	59.5
Gamma H (at %)	21.3	21.2	21.3	20.0	21.3
Simulated Distill.					
Initial boiling point (°C)	439	450	–	425	425
5% boiling point (°C)	536	–	–	504	532
10% boiling point (°C)	556	–	–	541	554
20% boiling point (°C)	583	–	–	574	581
SARA Analysis					
Saturates	16.3	16.3	–	16.8	16.3
Aromatics (wt %)	58.7	58.1	–	57.0	58.7
Resins (wt %)	18.9	19.6	–	20.9	18.9
Asphaltenes (wt %)	6.1	6.0	–	5.4	6.1

Table 3: Prediction of the properties of the saturates fraction from the Arabian Light vacuum residue

Saturates from the AL vacuum residue	Experimental data (Neurock, 1992)	Predicted (EM output)
Elementary Analysis		
Carbon (wt %)	85.82	86.76
Hydrogen (wt %)	14.02	13.24
Sulphur (wt %)	0.18	0
Nitrogen (wt %)	0.03	0
Oxygen (wt %)	0.73	0
Molecular weight (g/mol)	843	868
^{1}H NMR Analysis		
Aromatic H (at %)	0.5	0.0
Alpha H (at %)	1.7	0.0
Beta H (at %)	76.3	80.1
Gamma H (at %)	21.5	19.9

Table 4: Prediction of the properties of the aromatics fraction from the Arabian Light vacuum residue

Aromatics from the AL vacuum residue	Experimental data (Neurock, 1992)	Predicted (EM output)
Elementary Analysis		
Carbon (wt %)	83.45	84.82
Hydrogen (wt %)	10.56	9.64
Sulphur (wt %)	4.47	4.64
Nitrogen (wt %)	0.19	0.18
Oxygen (wt %)	0.70	0.72
Molecular Weight (g/mol)	928	972
^{1}H NMR Analysis		
Aromatic H (at %)	8.5	8.1
Alpha H (at %)	12.9	14.1
Beta H (at %)	57.1	55.4
Gamma H (at %)	21.6	22.4

Table 5: Prediction of the properties of the resins fraction from the Arabian Light vacuum residue

Resins from the AL vacuum residue	Experimental data (Neurock, 1992)	Predicted (EM output)
Elementary Analysis		
Carbon (wt %)	82.38	82.67
Hydrogen (wt %)	9.57	9.46
Sulphur (wt %)	5.02	4.96
Nitrogen (wt %)	0.78	0.86
Oxygen (wt %)	2.25	2.05
Molecular Weight (g/mol)	1278	1130
^{1}H NMR Analysis		
Aromatic H (at %)	8.3	8.5
Alpha H (at %)	15.7	13.2
Beta H (at %)	58.2	57.2
Gamma H (at %)	17.8	21.1

Table 6: Prediction of the properties of the asphaltenes fraction from the Arabian Light vacuum residue

Asphaltenes from the AL VR	Experimental data (Neurock, 1992)	Predicted (EM output)
Elementary Analysis		
Carbon (wt %)	84.06	84.71
Hydrogen (wt %)	7.59	8.94
Sulphur (wt %)	6.17	5.91
Nitrogen (wt %)	0.75	0.40
Oxygen (wt %)	1.43	0.04
Molecular Weight (g/mol)	1950	1370
^{1}H NMR Analysis		
Aromatic H (at %)	14.2	16.1
Alpha H (at %)	12.2	16.5

Beta H (at %)	44.9	51.9
Gamma H (at %)	28.8	15.5

Using the above-described distributions and molecule construction scheme, the SR step was employed to generate an initial set of 5000 molecules. The choice of the size of the set of molecules was a compromise between the computing time and the representativeness of the equimolar mixture. Table 2 also contains the calculated properties for this set of 5000 molecules after minimisation of an objective function that contained the elemental analysis, the average molecular weight, the ^{1}H NMR, the SARA analysis and the partial simulated distillation curve. The minimization was performed by means of an elitist genetic algorithm (De Jong, 1975; Goldberg, 1989; Goldberg et al., 1992), starting with a population of 1024 individuals. A ranking method was used to select the individuals for one-point crossover. On the off-spring, a single-point mutation was applied to each parameter, with a mutation probability of 0.02. Both the crossover and the mutation are carried out by means of floating point operators as described by Parker (1995). The minimisation was continued for 60 generations.

As seen from Table 2, the properties of the equimolar mixture of molecules obtained at the end of the SR step are already close to the corresponding experimental values. The elemental analysis was well predicted, although there was a surplus of oxygen. Looking at the ^{1}H NMR analysis, the calculated aromatic hydrogen content was too low and the aliphatic structures on the molecules are probably too large. Some differences also exist in the simulated distillation curve. Lastly, the recalculated SARA analysis correlates relatively well with the experimental data. More particularly, good results were obtained in the distribution of the molecules between the aromatic and resin families, for which no adjustable parameter was provided. In conclusion, the set of 5000 molecules obtained by SR can be considered as representative of an Arabian Light vacuum residue. Starting from this equimolar set of molecules, their molar fractions were then calculated during the EM step. The distribution of the molar fractions calculated by the EM method is shown in Fig. 4. The molar fractions range from 7.08×10^{-12} to 1.27×10^{-2}. The molecules with the lowest mole fractions appeared to be the heavy aromatics and resins. The final mixture properties are given in Table 2. The difference between these results and the experimental data is very small, and in most cases there is no difference at all. Only the estimated initial boiling point showed no improvement over the SR step. This is due to the fact that the light molecules created in the SR step have not been eliminated by the EM step. As can be seen from Table 2, the EM step of the two-step molecular reconstruction algorithm allowed to significantly reduce the deviations from the analytical results, without lessening the polydispersity and molecular diversity of the mixture (Fig. 5).

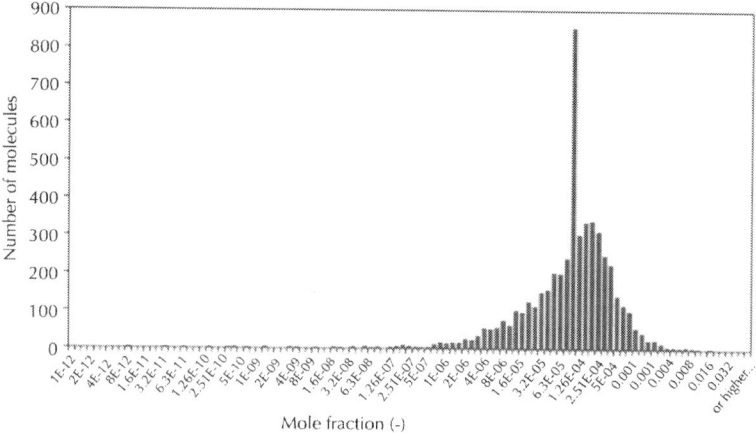

Figure 4: Distribution of the molar fractions.

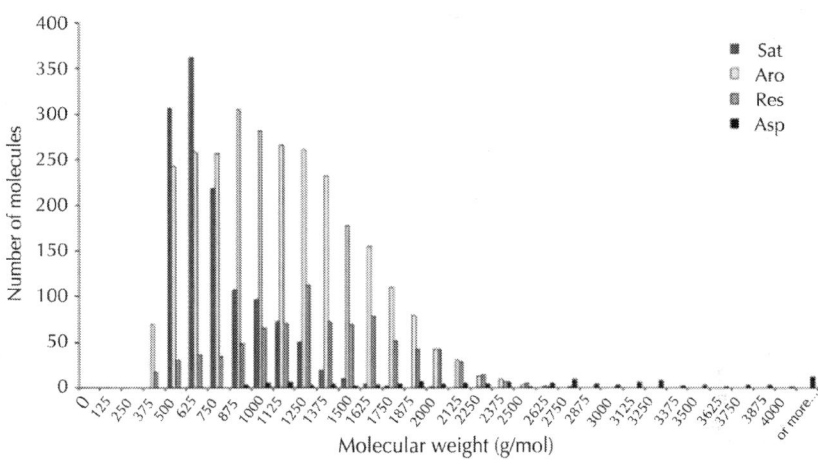

Figure 5: Molecular weight distribution by SARA family.

Up to this point, it was shown that the two-step molecular reconstruction algorithm is able to adjust the selected molecules and the mixture properties to an experimental set of data. We will now look into the predictive power of the method. As mentioned above, the 4 SARA fractions (saturates, aromatics, resins, and asphaltenes) were isolated by means of a preparative separation technique and subjected to molecular weight measurement, elementary analysis, and ^{1}H NMR analysis (Neurock, 1992; Trauth, 1993). It should be stressed that this information was not used during the reconstruction of the Arabian Light vacuum residue, and is therefore a real measure of the predictive power of the molecular reconstruction method.

Table 3, Table 4, Table 5 and Table 6 compare the properties of each of these fractions of the reconstructed mixture to the analytical data. For the saturates fraction, the molecular weight is well predicted. The elemental analysis shows that the prediction of hydrogen content of the saturates fraction is somewhat too low. This is probably due to the fact that the generated mixture contains too much naphthenes and lacks paraffins. As indicated by the experimental elemental and NMR analysis, the saturates contain small amounts of hetero-atoms and some aromatics. As the reconstruction scheme assumes that the paraffins and naphthenes only contain molecules with hydrogen and carbon, this cannot be predicted.

Both the aromatics and resins fractions are well predicted. For both fractions, almost the same overall hydrogen content is found, although that for the aromatics should be significantly higher. The nitrogen and oxygen contents are well predicted for both fractions. For the ^{1}H NMR analysis, the aromatic hydrogen content is also correctly predicted, but the distribution between the different types of aliphatic hydrogen atoms is somewhat skewed.

Regarding the asphaltenes, their molecular weight and their nitrogen and oxygen contents are underestimated. This is due to the fact that the EM step tends to attribute very low mole fractions to the heaviest asphaltenes, of the order of 10^{-10}. As these asphaltenes contain a large number of heteroatoms, their contents are underestimated and the predicted hydrogen and carbon contents are above the measured values. Finally, the prediction of the ^{1}H NMR analysis is relatively inaccurate in the case of the asphaltenes. The beta hydrogen content was highly overpredicted, whereas the predicted value was too low for the gamma hydrogen atoms. Based on their definition, it can be deduced that the lateral chains of the molecules were too long.

The generated set of molecules can now also be analysed in terms of molecular structures. Of course, the example molecules that are presented here are solely as an indication, and do not claim to be absolutely accurate. For the aromatics and resins families, Fig. 6 and Fig. 7 present sulphur-, nitrogen- and oxygen-containing molecules, whose properties are close to the average properties of their fraction. The resins are characterised by higher heteroatom contents, both in the polycyclic core or in the aliphatic chains. Moreover, the aromatic carbon/aliphatic carbon ratio is higher for resins molecules. These two findings are linked to the higher polarity of the resins. Finally, it must be remembered that the positioning of the chains, obtained by sampling of distribution 10, is entirely random. Hence, this results sometimes in improbable situations, in which the lateral side-chains are highly promiscuous.

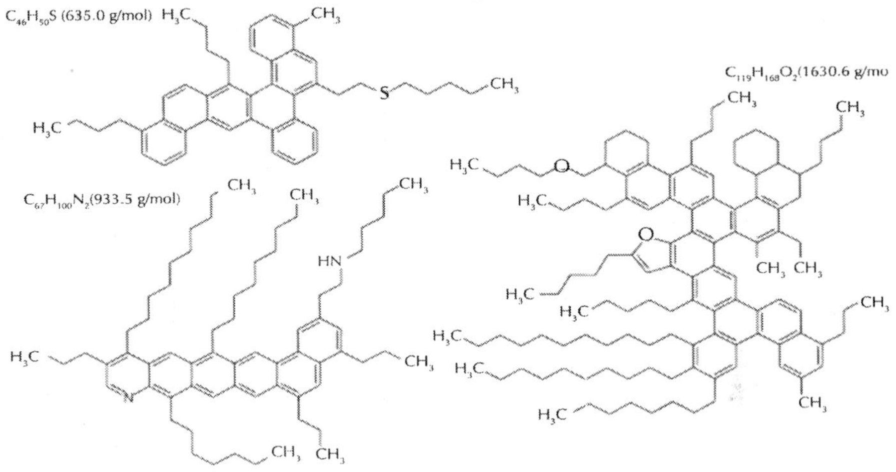

Figure 6: Sulphur-, nitrogen- and oxygen-containing molecules of the aromatics fraction.

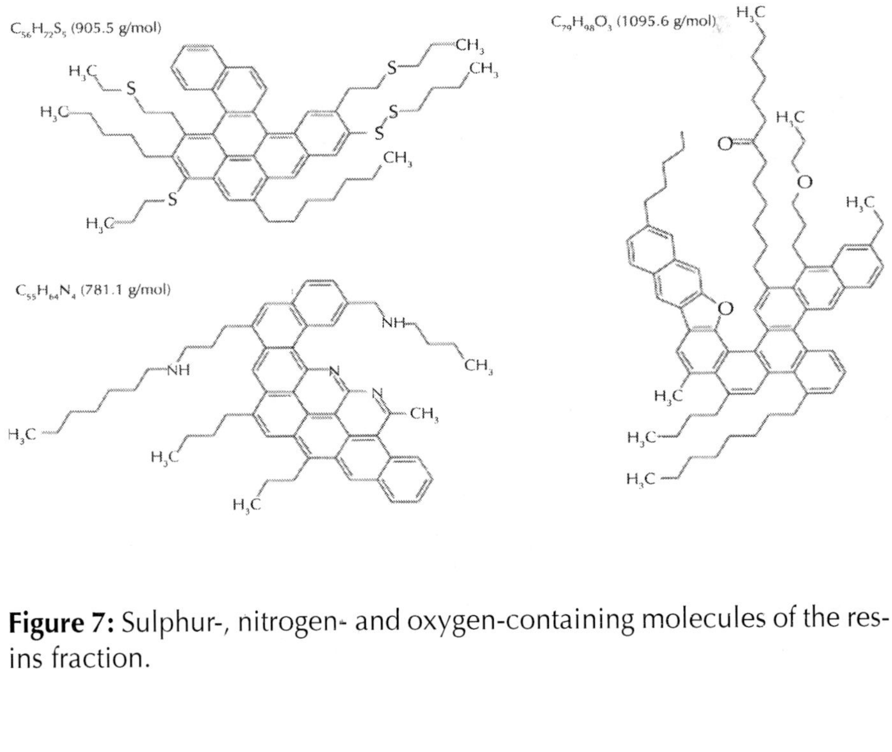

Figure 7: Sulphur-, nitrogen- and oxygen-containing molecules of the resins fraction.

Fig. 8 shows the molecular structure of a reconstructed asphaltene. Its representation only partly matches reality, due to the fact that the reconstruction scheme has imposed a number of constraints in order to minimise the number of distributions and parameters. Of course, the diversity of the structures that are present does not appear in the example. However, the model is close to the results obtained by Strausz et al. (1999) who anticipated a structure composed of slightly condensed polycyclic cores connected to a large number of lateral chains. With regard to the functional groups, these are mainly sulphurs and disulphurs, an ether bridge and an amine. They are also present on both the lateral chains and the chains linking the cores. In this example, the molecule selected contains no heteroatoms in its nuclei.

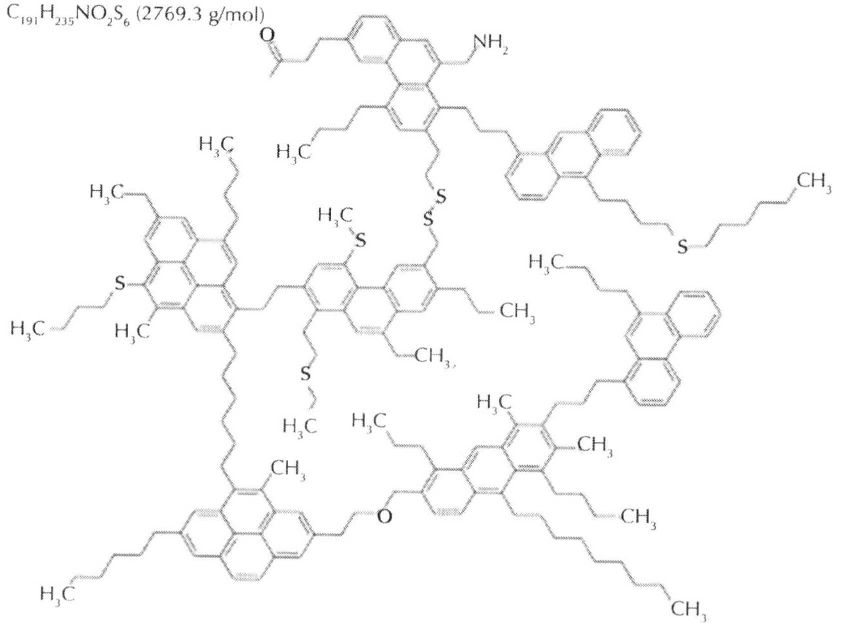

Figure 8: Representation of an asphaltene molecule for the Arabian Light vacuum residue.

CONCLUSIONS

To represent a petroleum fraction at the molecular level, a two-step molecular reconstruction method has been developed and was in this work applied to an Arabian Light vacuum residue.

In the first step of this molecular reconstruction algorithm, stochastic reconstruction was used to generate a large number of representative molecules, while the entropy maximisation method was employed in a second step to adjust their molar fractions so as to obtain a mixture of molecules, whose overall characteristics are very close to those provided by the analytical data of the residue fraction. As the available analyses from these heavy petroleum fractions are global in nature, the reconstruction problem was intrinsically only slightly constrained and was therefore characterised by large degrees of freedom during the construction of the chemical structures. Despite this, the two-step molecular reconstruction method allowed to generate a complex mixture of a large range of chemically feasible molecules, whose mixture properties correspond very closely to those of the Arabian Light vacuum residue.

To validate the composition of the reconstructed mixture and to assess the predictive power of the method, the molecular weight, elemental analysis and ^{1}H NMR signature of the four solubility classes (saturates, aromatics, resins and asphaltenes) were back-calculated and used as a validation against the analytical results obtained on these four fractions. Without using any additional data, the predicted characteristics can be considered good to very good for the saturates, the aromatics and the resins. For the asphaltenes, larger discrepancies were observed due to the fact that the entropy maximisation step tends to attribute very low mole fractions to the heaviest asphaltenes.

Finally, a few of the generated structures from the polydisperse mixture were presented to illustrate the type of compounds generated by the molecular reconstruction algorithm.

Overall, the two-step molecular reconstruction algorithm that was developed and adapted to residue fraction can be used to obtain a polydisperse molecular representation of these heavy petroleum products and was shown to be able to provide an estimate of missing analytical data. These mixtures can also be used as input to detailed kinetic models.

REFERENCES

1. Adam, F., Bertoncini, F., Thiébaut, D., Esnault, S., Espinat, D., Hennion, M.C., 2007. Towards Comprehensive Hydrocarbons Analysis of Middle Distillates by LC–GC×GC. Journal of Chromatographic Science 45 (10), 643–649.
2. Allen, D.T., Liguras, D.K., 1991. Structural models of catalytic cracking chemistry: a case study of a group contribution approach to lumped kinetic modeling. In: Sapre, A.V., Krambeck, F.J. (Eds.), Chemical Reactions in Complex Mixtures, Mobil Workshop. Van Nostrand Reinhold, New-York.
3. Ali, L.H., Al-Ghannam, K.A., Al-Rawi, J.M., 1990. Chemical structure of asphaltenes in heavy crude oils investigated by n.m.r. Fuel 69, 519–521.
4. Ali, F.A., Chaloum, N., Hauser, A., 2006. Structure representation of asphaltene GPC fractions derived from Kuwaiti residual oils. Energy & Fuels 20, 231–238.
5. Altgelt, K.H., Boduszynski, M.M., 1994. Composition and Analysis of Heavy Petroleum\ Fractions. Marcel Dekker, New York.
6. Artok, L., Su, Y., Hirose, Y., Hosokawa, M., Murata, S., Nomura, M., 1999. Structure and reactivity of petroleum-derived asphaltene. Energy & Fuels 13, 287–296.
7. Brown, J.K., Ladner, W.R., Sheppar, N., 1960. An analysis of the behavior of a class of genetic adaptive systems. Fuel 39, 79–86.

8. De Jong, K., 1975. An analysis of the behavior of a class of genetic adaptive systems. Ph.D. Thesis, University of Michigan.
9. Dickinson, E.M., 1980. Structural comparison of petroleum fractions using proton and 13C N.M.R. spectroscopy. Fuel 59 (5), 290–294.
10. Faulon, J.L., Vandenbroucke, M., Drappier, J.M., Behar, F., Romero, M., 1990. Modélisation des Structures Chimiques des Macromolécules Sédimentaires: le logiciel XMOL. Revue de l'Institut Français du Pétrole 45 (2), 161–180.
11. Fenistein, D., Barre, L., Broseta, D., Espinat, D., Livet, A., 1998. Viscosimetric and neutron scattering study of asphaltene aggregates in mixed toluene/heptane solvents. Langmuir 14, 1013–1020.
12. Gauthier, T., Danial-Fortain, P., Merdrignac, I., Guibard, I., Quoineaud, A.-A., 2008. Studies on the evolution of asphaltene structure during hydroconversion of petroleum residues. Catalysis Today 130, 429–438.
13. Goldberg, D.E., 1989. Genetic Algorithms in Search, Optimization and Machine Learning. Addison-Wesly, New York.
14. Goldberg, D.E., Deb, K., Clark, J.H., 1992. Genetic algorithms, noise, and the sizing of populations. Complex Systems 6, 333–362.
15. Groenzin, H., Mullins, O.C., 2000. Molecular size and structure of asphaltenes from various sources. Energy & Fuels 14, 677–684.
16. Hirsch, E., Altgelt, K.H., 1970. Integrated structural analysis. A method for the determination of average structural parameters of petroleum heavy ends. Analytical Chemistry 42 (12), 1330–1339.
17. Hudebine, D., Vera, C., Wahl, F., Verstraete, J., 2002. Molecular representation of hydrocarbon mixtures from overall petroleum analyses. In: 2002 A.I.Ch.E. Spring Meeting, New Orleans, LA, March 10–14, 2002, Paper 27a.

18. Hudebine, D., 2003. Reconstruction moléculaire de coupes pétrolières. Ph.D. Thesis, Ecole Nationale Supérieure de Lyon.
19. Hudebine, D., Verstraete, J.J., 2004. Molecular reconstruction of LCO gas oils from overall petroleum analyses. Chemical Engineering Science 59, 4755–4763.
20. Jaffe, S.B., Freund, H., Olmstead, W.N., 2005. Extension of structure-oriented lumping to vacuum residua. Industrial & Engineering Chemistry Research 44, 9840–9852.
21. Khorasheh, F., Khaledi, R., Gray, M.R., 1998. Computer generation of representative molecules for heavy hydrocarbon mixtures. Fuel 77 (4), 241–253.
22. Miyabayashi, K., Naito, Y., Tsujimoto, K., Miyake, M., 1995. Structure characterization of petroleum residuum components fractionated by high vacuum short-path distillation (DISTACT) and gel permeation chromatography (GPC). In: ACS 210th National Meeting (Chicago August 20–25, 1995), ACS Division of Fuel Chemistry Preprints, 40, 3, 497–503.
23. Mullins, C., Sheu, E.Y., 1999. Structures and Dynamics of Asphaltenes. Springer, Berlin. Mullins, C., Sheu, E.Y., Hammami, A., Marshall, A.G., 2006. Asphaltenes, Heavy Oils, and Petroleomics. Springer, Berlin.
24. Murgich, J., 2003. Molecular simulation and the aggregation of the heavy fractions in crude oils. Molecular Simulation 29, 451–461.
25. Murgich, J., Abanero, A., Strausz, O.P., 1987. Molecular recognition in aggregates formed by asphaltene and resin molecules from athabasca oil sands. Energy & Fuels 1 (5), 381–386.
26. Neurock, M., 1992. A computational chemical reaction engineering analysis of complex heavy hydrocarbon reaction systems. Ph.D. of the University of Delaware.
27. Neurock, M., Nigam, A., Trauth, D.M., Klein, M.T., 1994. Molecular representation of complex hydrocarbon feedstocks through efficient characterization and stochastic algorithms. Chemical Engineering Science 49 (24), 4153–4177.

28. Parker, J.R., 1995. Genetic operators on floating point parameters. Department of Computer Science Research Report #95/564/16. University of Calgary. http://dspace.ucalgary.ca/bitstream/1880/46436/2/1995-564-16.pdf.
29. Quann, R.J., Jaffe, S.B., 1992. Structure-oriented lumping: describing the chemistry of complex hydrocarbon mixtures. Industrial & Engineering Chemistry Research 31, 2483–2497.
30. Quann, R.J., Jaffe, S.B., 1996. Building useful models of complex reaction systems in petroleum refining. Chemical Engineering Science 51 (10), 1615–1635.
31. Revellin, N., Dulot, H., López-García, C., Baco, F., 2005. Specific nitrogen boiling point profiles of vacuum gas oils. Energy & Fuels 19, 2438–2444.
32. Revellin, N., 2006. Modélisation de l'hydrotraitement des distillats sous vide. Ph.D. Thesis, Ecole Normale Supérieure de Lyon.
33. Sato, S., 1997. The development of a support program for the analysis of average molecular structures by personal computer. Sekiyu Gakkaishi 40 (1), 46–51.
34. Sheu, E.Y., 2002. Petroleum asphaltene-properties, characterization, and issues. Energy & Fuels 16, 74–82.
35. Sheu, E.Y., Mullins, O.C., 1996. Asphaltenes. Springer, Berlin.
36. Speight, J.G., 1999. The Chemistry and Technology of Petroleum, third ed. Marcel Dekker Inc., New York.
37. Strausz, O.P., Mojelsky, T.W., Faraji, F., Lown, E.M., 1999. Additional structural details on athabasca asphaltene and ramifications. Energy & Fuels 13, 207–227.
38. Suzuki, T., Itoh, M., Takegami, Y., Watanabe, Y., 1982. Chemical structure of tar-sand bitumens by 13C and 1H n.m.r. spectroscopic methods. Fuel 61, 402–410.
39. Takegami, Y., Watanabe, Y., Suzuki, T., Mitsudo, T., Itoh, M., 1980. Structural investigation on column-chromatographed vacuum residues of various petroleum crudes by 13C nuclear magnetic resonance spectroscopy. Fuel 59, 253–259.

40. Trauth, D.M., 1993. Structure of Complex Mixtures through Characterization, Reaction, and Modeling. Ph.D. of the University of Delaware.
41. Trauth, D.M., Stark, S.M., Petti, T.F., Neurock, M., Klein, M.T., 1994. Representation of the molecular structure of petroleum resid through characterization and Monte Carlo modeling. Energy & Fuels 8 (3), 576–580.
42. Van Geem, K.M., Hudebine, D., Reyniers, M.F., Wahl, F., Verstraete, J.J., Marin, G.B., 2007. Molecular reconstruction of naphtha steam cracking feedstocks based on commercial indices. Computers & Chemical Engineering 31, 1020–1034.
43. Verstraete, J.J., Revellin, N., Dulot, H., Hudebine, D., 2004. Molecular reconstruction of vacuum gas oils. Preprint Paper—American Chemical Society, Division of Fuel Chemistry 49 (1), 20–21.
44. Zhao, S., Xu, Z., Xu, C., Chung, K.H., Wang, R., 2005. Systematic characterization of petroleum residua based on SFEF. Fuel 84 (6), 635–645

Citations

CHAPTER 1

Christine Cleetus, Shijo Thomas, and Soney Varghese, "Synthesis of Petroleum-Based Fuel from Waste Plastics and Performance Analysis in a CI Engine," Journal of Energy, vol. 2013, Article ID 608797, 10 pages, 2013. doi:10.1155/2013/608797.

CHAPTER 2

Muhammad Imran Ahmad, Nan Zhang, Megan Jobson, Molecular components-based representation of petroleum fractions, Chemical Engineering Research and Design, Volume 89, Issue 4, April 2011,

Pages 410-420, ISSN 0263-8762, http://dx.doi.org/10.1016/j.cherd.2010.07.016.

CHAPTER 3

Damien Hudebine, Jan J. Verstraete, Molecular reconstruction of LCO gasoils from overall petroleum analyses, Chemical Engineering Science, Volume 59, Issues 22–23, November–December 2004, Pages 4755-4763, ISSN 0009-2509, doi.org/10.1016/j.ces.2004.09.019.

CHAPTER 4

L. Zuñiga Liñan, N.M. Nascimento Lima, F. Manenti, M.R. Wolf Maciel, R. Maciel Filho, L.C. Medina, Experimental campaign, modeling, and sensitivity analysis for the molecular distillation of petroleum residues 673.15 K+, Chemical Engineering Research and Design, Volume 90, Issue 2, February 2012, Pages 243-258, ISSN 0263-8762, http://dx.doi.org/10.1016/j.cherd.2011.07.001.

CHAPTER 5

Lu Cheng, Lanlan Ye, Dejun Sun, Tao Wu, Yujiang Li, Removal of petroleum sulfonate from aqueous solution by hydroxide precipitates generated from leaching solution of white mud, Chemical Engineering Journal, Volume 264, 15 March 2015, Pages 672-680, ISSN 1385-8947, http://dx.doi.org/10.1016/j.cej.2014.12.003.

CHAPTER 6

Iranildes D. Santos, Márcia Dezotti, Achilles J.B. Dutra, Electrochemical treatment of effluents from petroleum industry using a Ti/

RuO2 anode, Chemical Engineering Journal, Volume 226, 15 June 2013, Pages 293-299, ISSN 1385-8947, http://dx.doi.org/10.1016/j.cej.2013.04.080.

CHAPTER 7

Reuben N. Okparanma, Frederic Coulon, Abdul M. Mouazen, Analysis of petroleum-contaminated soils by diffuse reflectance spectroscopy and sequential ultrasonic solvent extraction–gas chromatography, Environmental Pollution, Volume 184, January 2014, Pages 298-305, ISSN 0269-7491, http://dx.doi.org/10.1016/j.envpol.2013.08.039.

CHAPTER 8

J.J. Verstraete, Ph. Schnongs, H. Dulot, D. Hudebine, Molecular reconstruction of heavy petroleum residue fractions, Chemical Engineering Science, Volume 65, Issue 1, 1 January 2010, Pages 304-312, ISSN 0009-2509, http://dx.doi.org/10.1016/j.ces.2009.08.033.

Index

A

Algebraic Equations (AEs) 107
Atmospheric Equivalent Boiling Point (AEBP) 88
Average Absolute Deviation (AAD) 122

C

Compression ignition (CI) 2

D

Differential Scanning Calorimeter (DSC) 101
Diffuse reflectance spectra 192, 198, 217

E

Energy dispersive X-ray fluorescence (EDXRF) 171
Energy necessary 168, 183
Entropy maximization (EM) 231
Entropy maximization\" (EM) 226
Experimental factorial design (EFD) 95

F

Field ionization detector (FID) 100
Fourier transform infrared 193, 219
Freshly generated hydroxide precipitates (FGHPs) 134, 136

G

Gas Chromatography (GC) 89
Gas chromatography-mass spectrometry (GC-MS) 14
Gas chromatography–mass spectrometry (GC–MS) 192
Gas chromatography with mass spectrometry (GC/MS) 63

H

High performance liquid chromatography (HPLC) 224
High-Temperature Simulated Distillation (HTDS) 89

I

Initial Boiling Point (IBP) 89
Isoelectric point (IEP) 139

L

Leaching solution of white mud (LSWM) 134, 136
Light cycle oil (LCO) 63, 226
Limit of quantitation (LOQ) 196

M

Molecular distillation 84, 109, 128, 132
Molecular type and homologous series (MTHS) 28, 33

N

Nuclear magnetic resonance (NMR) 31, 224

P

Partial least-squares regression analysis 191, 202
Petroleum sulfonate (PS) 134, 135
Plug-flow reactor (PFR) 170

Polycyclic aromatic hydrocarbons 192, 204, 205, 208, 209, 215, 217, 219

R

Rare earth metal-exchanged Y-type (REY), 7
Ratio of prediction deviation 192, 200
Reconstruction by entropy maximization (REM) 79
Reconstruction by entropy maximization\" (REM). 63
Root-mean-square error (RMSE) 200

S

Size exclusion chromatography (SEC) 224
Soil spectra 199
Spectral reflectance 191, 199, 209, 210
Spectrophotometer 192, 198, 216, 219
Spectroscopy 193, 194, 207, 208, 213, 214, 216, 217, 218, 219
Statistical distributions (SR) 63
Stochastic reconstruction (SR) 79, 226, 227
Stochastic reconstruction\" (SR) 225
Stochastic reconstruction\" (SR). 63
Structure oriented lumping\" (SOL) 225
Structure-oriented lumping (SOL) 33

T

The true-boiling point (TBP) 38
Thin layer chromatography (TLC) 100

U

United States Department of Agriculture (USDA) 195

V

VPO (Vapor Pressure Osmometry) 99